Napkin Folding

DR. Raj Kumar Singh
D.H.C.T Mgmt. (W.B)
P.G.D.B Mgmt. (U.P)
C.F.N. (Delhi)
MHCIMA (London)
Ph.D (Management)
'SAMS' Institute of Hotel Management,
Varanasi

THE HOSPITALITY PRESS
DELHI (INDIA)

Published by:
THE HOSPITALITY PRESS
7360-B, PREM NAGAR, GALI NO. 1,
SHAKTI NAGAR, DELHI-110007
Phone : 9013629093, 011-23824937;
e-mail: thehospitalitypress@gmail.com

(Reprint) 2010
Revised Edition: 2026
ISBN: 978-81-820401-8-2

SOLE DISTRIBUTOR
JINDAL BOOK SERVICES
LG-4A, Ganpati Bhawan
4675-B/21, Ansari Road,
Darya Ganj, Delhi - 110002

Typesetting and Jacket Designed by:
S.K. Educational Books

Printed by:
G.K.Fineart Press, Delhi, India

Preface

Napkin folding is much more than a decorative practice; it is an integral component of dining culture and hospitality service. By combining function, aesthetics, history, and cultural meaning, napkin folding enhances the overall dining experience and reflects the values of care and professionalism. Whether in a luxury hotel, a festive banquet, or a home dining setting, the art of napkin folding demonstrates how small details can leave a lasting impression on guests.

Napkin folding is an essential aspect of table presentation that blends functionality with aesthetic expression. It transforms a simple piece of fabric or paper into a visual element that enhances the overall dining experience. Commonly associated with fine dining, banquets, and special occasions, napkin folding has gradually become a recognized skill within hospitality, event management, and even home entertaining.

Historically, napkin folding developed alongside the evolution of dining customs. In early communal dining settings, cloths were primarily used for practical purposes such as wiping hands or protecting garments. As table etiquette became more refined, particularly in European courts and aristocratic households, napkins gained a symbolic role. Elaborate folds began to appear as demonstrations of sophistication, craftsmanship, and hospitality. Over time, these practices spread globally, adapting to local traditions, materials, and dining styles. Today, napkin folding carries both historical significance and modern relevance.

From a functional perspective, napkin folding serves practical needs while maintaining order and cleanliness on the table. A well-folded napkin signals readiness for service and provides guests with easy access during the meal. Different folds can indicate the type of service being offered, such as formal dining, buffet arrangements, or casual settings. In professional environments like hotels and restaurants, standardized folds help maintain consistency and efficiency, ensuring that table setups meet established service standards.

Aesthetically, napkin folding contributes to the visual harmony of table settings. The shape, symmetry, and placement of a folded napkin complement tableware, linens, and décor. Seasonal and thematic folds allow hosts and hospitality professionals to align presentation with specific occasions, festivals, or events. For example, floral or fan-shaped folds may be used in celebratory settings, while simple geometric folds suit formal or corporate environments. Through thoughtful design, napkin folding becomes a subtle yet powerful form of visual communication. Material selection is another important aspect of napkin folding. Cloth napkins, often made of cotton or linen, are preferred in formal and sustainable settings due to their durability and refined appearance. Paper napkins are commonly used in casual dining and large events for convenience and cost efficiency. The choice of material affects the complexity and stability of folds, as stiffer fabrics hold shapes more effectively. Skilled folding requires awareness of material properties to achieve both visual appeal and practical usability.

In contemporary hospitality and event management, napkin folding is closely linked to branding and guest experience. Unique or signature folds can become part of a venue's identity, reinforcing brand image and service quality. Training in napkin folding is often included in hospitality education, emphasizing precision, consistency, and creativity. As sustainability becomes a growing concern, reusable napkins and simple yet elegant folds are gaining prominence, balancing aesthetics with environmental responsibility.

–DR. Raj Kumar Singh

Contents

1

Introduction to Napkin Folding as an Applied Art

Conceptual Foundations of Napkin Folding

The historical significance of dining etiquette lies in its ability to reveal how societies have understood order, identity, and values across time. From ancient rituals to modern global interactions, dining practices have served as mirrors of social structure and cultural priorities.

By studying the evolution of dining etiquette, one gains a deeper appreciation of how everyday behaviors are shaped by historical forces and how they continue to influence social interaction. Dining etiquette remains a living tradition, connecting past and present through shared practices that define how people come together to eat, communicate, and express respect within their communities.

How to Fold a Napkin into the Shape of a Bouquet: Step-by-Step Guide with Photos and Description

On our portal, you will find many tutorials with photos and videos on napkin folding. Before you start folding a napkin into the shape of a bouquet, I recommend watching the video. Then you can follow the step-by-step guide.

We have photographed every action so you can easily follow along. Our tip: Do not skimp on the quality of the napkins. A cheap white napkin does not hold its shape well and does not unfold the magic that transforms a regular meal into a festive dinner.

Step 1: Take two napkins of different sizes and a thin string or narrow ribbon to tie the napkin. First, fold the larger napkin diagonally.

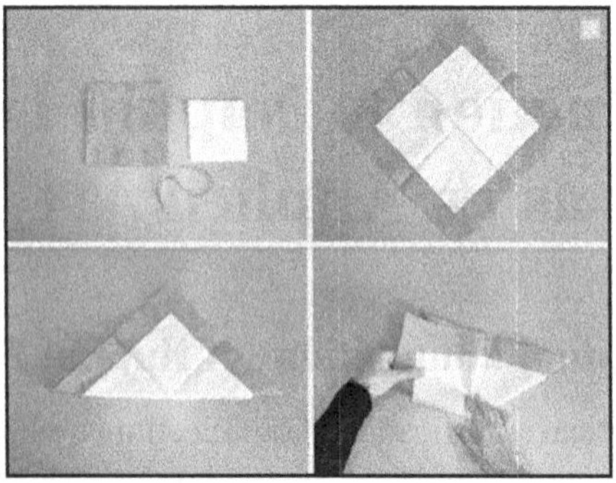

Step 2: Fold two corners inwards.

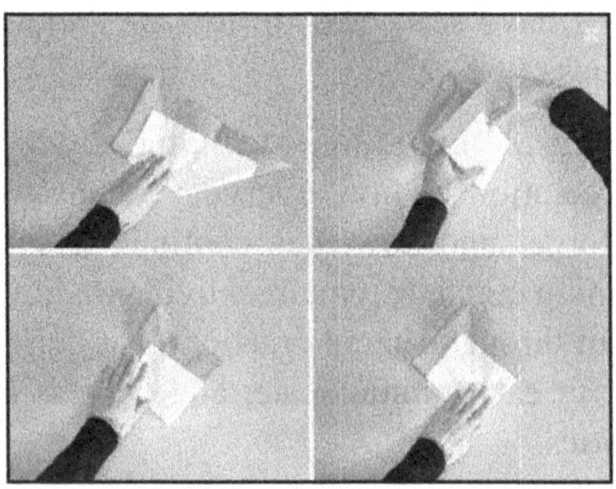

Step 3: Now fold all four corners inwards. Repeat this process all around until you get a round shape.

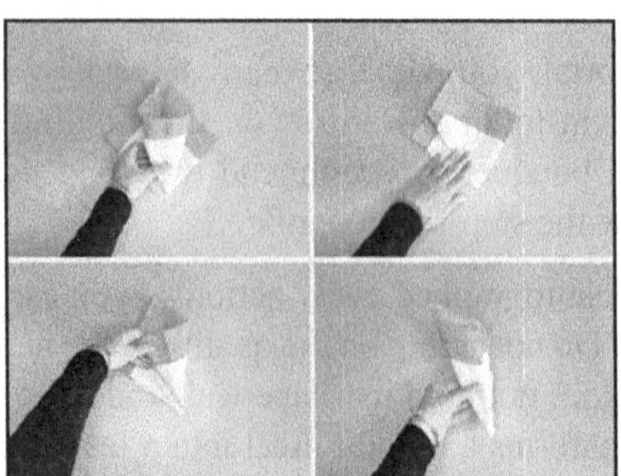

Step 4: Leave about two centimeters free at the bottom and tie the napkin together with a knot. Your napkin bouquet is ready!

Folding a Napkin into the Shape of a Bouquet: Conclusion and Thank You

We created this guide specifically for you and would be delighted if you try folding your napkin following our instructions. It would be wonderful if your celebration or cozy evening becomes even more beautiful as a result, and you enjoy it. We are grateful that you took the time to read and try our guide.

Geometric Blueprint

The primary conceptual foundation of napkin folding is rooted in the pure and unyielding principles of geometry. Every napkin fold begins with a simple plane, most often a square. This shape is the blank canvas, a field of infinite potential governed by mathematical rules. The entire art form is a dialogue with this initial geometry.

The fundamental actions—the valley fold and the mountain fold—are the basic vocabulary of a language that can create staggering complexity from utter simplicity. These are the same foundational elements seen in the more celebrated art of origami, and the principles are directly transferable. A crease is not just a line; it is a decision, an irreversible alteration of the plane that introduces structure and memory into the fabric. Each fold is a geometric transformation. A simple diagonal fold creates two right-angled triangles from a square, introducing symmetry and bisecting the original form. A book fold creates two rectangles, exploring a different kind of division. As these simple folds are layered, the process becomes a cumulative exercise in spatial reasoning.

The creator must visualize how each new crease will interact with the previous ones, how flat surfaces will become points, and how edges will align to create new patterns. The resulting forms, whether the sharp, architectural lines of a "Pyramid" or the repeating angles of a "Fan," are direct products of this geometric syntax. The elegance of a finished napkin fold is therefore a testament to the mathematical precision and ordered thinking required to bring it into being.

Sculpture in Miniature

Fig: Fiolding of Napkins in Minature forms

Beyond its geometric purity, napkin folding operates as a form of miniature sculpture. The process is a fundamental sculptural act: the transformation of a two-dimensional object into a three-dimensional form that occupies space and interacts with light. Once a napkin is folded, it possesses volume, structure, and presence. It is no longer just a flat piece of cloth; it is an object with a distinct silhouette, with faces, ridges, and hollows. The material itself—the weight of starched linen versus the crispness of thick paper—

becomes the medium, influencing the final texture and stability of the piece. A starched linen can hold a sharp, architectural crease, lending itself to formal, towering structures like the "Bishop's Mitre," which stands with a quiet authority on the plate.

This sculptural quality is defined by the interplay of form and negative space. A "Lotus" or "Rose" fold does not just create the shape of a flower; it also creates a captivating interior space and casts intricate shadows that add depth and complexity. The arrangement of folds directs the eye, creating rhythm and movement.

The repetitive pleats of a "Fan" create a visual cadence, while the spiraling form of a "Rosebud" draws the viewer's gaze inward towards its center. In this context, the napkin ceases to be a functional object and becomes a temporary installation on the landscape of the dinner plate. It commands attention and asserts an aesthetic identity, proving that the principles of sculpture—form, mass, and the manipulation of space—can be explored and expressed even at the most intimate scale.

Language of Welcome and Occasion

This sculptural quality, in turn, serves a deeper purpose as a medium of non-verbal communication. A folded napkin is one of the first things a guest interacts with at a table, and its form communicates a wealth of information before a single word is spoken. It is a silent host, setting the tone for the event and conveying the host's intentions.

A simple, casual fold, like a neatly placed rectangle or triangle, suggests a relaxed, informal atmosphere. It says, "Welcome, be comfortable." In contrast, an elaborate, complex fold signifies a special occasion. It speaks of effort, care, and a desire to create a memorable and elevated experience. The presence of a "Fleur-de-Lis" or an intricate "Bird of Paradise" communicates that this meal is a celebration, an event worthy of pomp and artistry. This language is understood intuitively across cultures. The investment of time and skill required to create a complex fold is a universally recognized gesture of respect and hospitality. It tells the guest that they are valued and that their experience matters. The choice of fold can also be thematic, adding another layer to the narrative of the meal. A "Sailboat" fold at a seaside luncheon or a "Christmas Tree" fold during the holidays reinforces the theme and enhances the overall aesthetic coherence of the event. In this way, the napkin becomes a key part of the semiotics of dining, a powerful symbol that helps define the social context and emotional register of the shared meal.

Poetics of Impermanence

Perhaps the most profound conceptual foundation of napkin folding is its inherent ephemerality. Unlike a painting or a marble sculpture, a folded napkin is created with the full knowledge of its own impending destruction. Its ultimate purpose is to be undone. The guest must unfold it, destroying the carefully constructed form in order to use it for its functional purpose. This places napkin folding in the category of ephemeral arts, alongside sand mandalas, ice sculptures, and performance art. The beauty and meaning of the object are inextricably linked to its transient nature.

This paradox—of investing skill and creativity into something so temporary—is what gives the act its poetic weight. The value is not solely in the finished product, but in the gesture of creation itself. The act of folding is an act of care, a moment of focused intention dedicated to the pleasure of another. The subsequent unfolding is not an act of destruction but a completion of the cycle, an acceptance of the gift.

The beauty of the folded napkin exists in a fleeting moment, to be appreciated and then relinquished. This impermanence encourages a state of mindfulness, prompting both the creator and the observer to appreciate the beauty of the present moment. It is a testament to the human desire to find order, beauty, and meaning in the most ordinary of objects, transforming a simple meal into a moment of transient, deliberate art.

Historical Significance in Dining Etiquette

Ancient Civilizations and the Origins of Dining Etiquette

The roots of dining etiquette can be traced back to ancient civilizations where meals were closely tied to ritual, religion, and social structure. In ancient Egypt, dining was often a hierarchical activity, with seating arrangements reflecting social rank and status. The act of sharing food was associated with hospitality and divine favor, and proper conduct during meals was considered a sign of moral discipline.

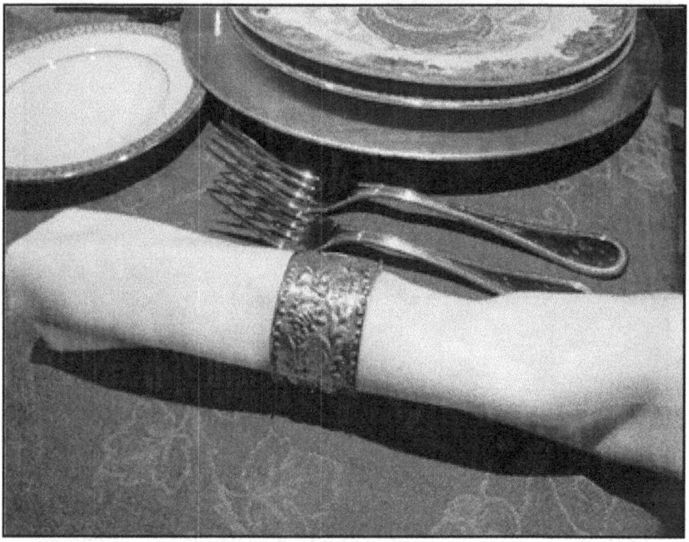

Fig: Use of Napkins in Ancient Times

Similarly, in ancient Greece, symposiums were not merely social gatherings but structured events with expectations regarding speech, behavior, and moderation. Dining etiquette emphasized self-control and intellectual engagement, reinforcing ideals of citizenship and philosophical balance. In ancient Rome, elaborate banquets highlighted power and wealth, while strict conventions governed posture, dress, and interaction, illustrating how etiquette functioned as a tool for displaying authority and reinforcing class divisions.

Napkins have been recorded as being in use, from the times of ancient Roman Empire and prior to then, in ancient Greece. References to word napkin dates back to 1384 AD. It is believed that the first napkin was a lump of dough the Spartans called "apomagdalie." It was a mixture that was cut into small pieces, rolled and kneeded at the table, that led to using sliced bread to wipe the hands.

Fig: Last Supper by Dieric Bouts Depicting the use of Napkins

In Roman antiquity, napkins known as "sudaria" and "mappae" were made in both small and large lengths. The sudarium, Latin for "handkerchief," was a pocket-size fabric earned to blot the brow during meals taken in the warm Mediterranean climate. The mappa was a larger cloth spread over the edge of the couch as protection from food taken in a reclining position. The fabric was also used to blot the lips. Each guest brought his own large mappa "napkin" and they became the first known "doggy bags" as guests leaving a

Roman dinner had their mappa filled with leftover delicacies from the enormous feasts to take home. With the fall of Roman Empire, napkins disappeared from the dining table.

Napkins returned to adorn the dining table many centuries later and the classic painting Last Supper from 1464-1467 AD by Dieric Bouts depicts the use of napkins on the dinner table. The use of the napkin in Europe began in 1400 on the tables of royalty, where they started to use napkins made from warm or even perfumed cloth. By the 16th century, napkins were part of rich dining experience and came in many sizes, known by various names like diaper, serviette, touaille (towel) depending on the size and intended use.

The art of napkin-folding in Europe began in Italy. The most popular, were those that were resembling animals. The folded napkins, or "serviettes", caught the eyes of all fashionable British, after the Restoration. Samuel Pepys wrote in 1669 in his diary, of a grand dinner held. The day prior, he wrote of having a caterer come to lay the table cloth and fold the napkins. 17th century saw the use of big sized napkins measuring 35" by 45" inches to help accommodate the needs of eating with bare hands instead of spoons and forks. The size reduced when forks and spoons were accepted as part of regular dining experience in major parts of Europe in the 18th century including Great Britain.

The 17th century also saw French come up with elaborate rules for nobility class which included instructions on napkin usage, a predecessor to modern day napkin etiquette, including the one which instructs the guests to not use the napkin for wiping the face or clean teeth or worse, rub one's nose.

"The French court imposed elaborate codes of etiquette on the aristocracy, among them the way to use a napkin, when to use it, and how far to unfold it in the lap. A French treatise dating from 1729 stated that "It is ungentlemanly to use a napkin for wiping the face or scraping the teeth, and a most vulgar error to wipe one's nose with it." And a rule of decorum from the same year laid out the protocol: "The person of highest rank in the company should unfold his napkin first, all others waiting till he has done so before they unfold theirs. When all of those present are social equals, all unfold together, with no ceremony.

"Fashionable men of the time wore stiffly starched ruffled collars, a style protected while dining with a napkin tied around the neck. Hence the expression "to make ends meet." When shirts with lace fronts came into vogue, napkins were tucked into the neck or buttonhole or were attached with a pin. In 1774, a French treatise declared, "the napkin covered the front of the body down to the knees, starting from below the collar and not tucked into said collar."– Margaret Visser

Fig: Napkin Folded into a Kimono with the Chopsticks Tucked in for the Diner to Pull Out for Use

Napkin rings or "serviette holders" began in the Napoleonic period in Europe. The rings were used mainly by upper class families as a way to identify which napkin belonged to what family member. At the beginning of the 1800s, napkins became part of the bourgeois lifestyle, mostly to protect the sumptuous dresses of the period during meals. The art really took off around 1880 with the incoming prosperity of the Industrial Revolution. One etiquette book advised "one should, if possible, avoid taking the bones into your mouth and when removing any, should shield your mouth with your napkin."

In 1860's "The Ladies' Book of Etiquette and Manual of Politeness" by Florence Hartley, Mrs. Hartley recommends one bring a pincushion to a dinner party for use with the napkin: "Sit gracefully at the table; neither so close as to make your movements awkward, nor so far away as to drag your food over your dress before it reaches your mouth. It is well to carry in your pocket a small pincushion, and, having unfolded your napkin, to pin it at the belt. You may do this quietly, without its being perceived, and you will thus really save your dress. If the napkin is merely laid open upon your lap, it will be very apt to slip down, if your dress is of silk or satin, and you risk the chance of appearing again in the drawing-room with the front of your dress soiled or greased."– Florence Hartley

And Arthur Martine's book of etiquette from the same period, gave advice on napkins and finger-bowls: "Finger-glasses are generally handed round as soon as the viands are removed, but they are intended merely to wet the fingers and around the mouth. When the finger-glasses are passed, wet your fingers in them and then wipe them upon your napkin. The habit of rinsing the mouth at table is a disgusting piece of indelicacy, which is never practiced by any well-bred person.Upon leaving the table, lay your napkin beside your plate, but do not fold it."

In 1895, George Rippey Stewart lamented the British and their habits on the use of napkins: "It is a matter of regret that table napkins are not considered indispensable in England; for, with all our boasted refinement, they are far from being general. The comfort of napkins at dinner is too obvious to require comment, whilst the expense can hardly be urged as an objection. If there be not any napkins, a man has no alternative but to use the table-cloth, unless (as many do) he prefers his pocket handkerchief— a usage sufficiently disagreeable."

In 1911, Agnes Morton, in her book "Etiquette," wrote about napkins: "Napkin rings are discarded by many who hold that a napkin should be used but once, and must be re-laundried before reappearing on the table.

Practically, such a fastidious use of table linen would exhaust most linen supplies, and overcrowd the laundry. The neat use of a napkin renders this extreme nicety superfluous as a rule of home dining, Care should certainly be taken to remove all soiled table linen. Nothing is more disgusting than a dirty napkin, but the snowy linen that comes spotless through one using may, with propriety, be retained in the ring to be used several times. This, of course, refers to every-day dining at home. On formal occasions no napkin rings appear on the table; the napkins are always fresh, and used for that time only. At the close of the dinner, they are left carelessly on the table; not rolled or folded in any orderly shape.

Fig: Origami Flower Folded Napkin in the Plate for the First Course

Small fringed napkins of different colors are used with a dessert of fruits. Fancy doylies of fine linen embroidered with silk are sometimes brought in with the finger-bowls; but these are not for utility, the dinner napkin doing service, while the embroidered 'fancy,' a dainty bit of effect to the table decoration.

Modern Day Napkin Etiquette

- Paper napkins can be flimsy, so it is okay to use more than one. In a pinch, a paper towel can be used as a napkin.
- Napkins are never a part of your clothing so do not use your shirt sleeve or shirt tail to wipe your mouth with.
- Napkins can never be dipped in your water glass to then wipe your face with, once you are old enough to eat on your own.
- Tablecloths are no longer napkins. They were at one time in history, but that was very long ago.
- Your wrists and the backs of your hands are not napkins.
- A napkin stays on your lap the entire time you are seated to eat. If you have to leave the table for a moment and others are still eating, your napkin goes on your seat until you return.
- Napkins do not go back on to the table until everyone at the table is finished eating.
- Paper napkins can be crumpled and put on your plate when you are done, if the plate will be thrown away, as well. Cloth napkins go beside your plate, to the left.
- Unless you are part of the family, do not put your napkin back into a napkin ring unless you have been advised to do so by your host or hostess.

Medieval Europe and the Moral Dimensions of Dining

During the medieval period, dining etiquette became closely linked with religious values and moral instruction. Meals were often communal, particularly in monasteries and noble households, where strict codes governed silence, prayer, and conduct. Eating was viewed as an act that required restraint, as excessive indulgence was associated with sin.

Table manners, such as eating modest portions and avoiding waste, reflected Christian teachings on humility and self-discipline. The introduction of manuals on proper behavior, often written for young nobles, marked a significant moment in the formalization of dining etiquette. These texts emphasized respect for elders, cleanliness, and decorum, demonstrating how dining practices were used to cultivate moral character and social responsibility.

Renaissance Influence and the Refinement of Table Manners

The Renaissance period brought a renewed focus on refinement, education, and humanist ideals, which significantly influenced dining etiquette. As courts became centers of cultural life, dining evolved into a sophisticated social performance. The use of individual utensils, including forks, gradually became more common, reflecting changing attitudes toward hygiene and personal space. Table manners were increasingly associated with civility and social polish, distinguishing the educated elite from the common population. Etiquette manuals expanded in scope, offering detailed guidance on behavior, conversation, and posture at the table. Dining etiquette during this era symbolized cultural advancement and intellectual refinement, reinforcing the idea that proper conduct was an essential marker of social identity.

Dining Etiquette in Asian Historical Traditions

In many Asian cultures, dining etiquette developed in close connection with philosophical and ethical systems. In China, Confucian principles played a central role in shaping table manners, emphasizing respect for elders, harmony, and hierarchical order. Seating arrangements, serving practices, and even the pace of eating were guided by deeply ingrained cultural values. Meals were seen as opportunities to reinforce family bonds and social obligations. Similarly, in Japan, dining etiquette evolved with an emphasis on simplicity, mindfulness, and respect for food. The aesthetic presentation of meals and rituals such as the tea ceremony reflected broader cultural ideals of balance and discipline. These traditions illustrate how dining etiquette functioned as an extension of moral philosophy and social harmony.

Fig: Chinese Dining Etiquette Survival

Colonial Encounters and the Global Exchange of Etiquette

The age of exploration and colonial expansion led to significant exchanges in dining customs across cultures. European colonial powers introduced their table manners to colonized regions, often viewing their own practices as symbols of superiority and modernity. At the same time, colonial encounters influenced European dining habits through the introduction of new foods, spices, and serving styles. Dining etiquette became a site of cultural negotiation, where local traditions blended with foreign influences. In many societies, adopting European dining manners became associated with social mobility and education, highlighting how etiquette was used as a marker of cultural assimilation and power dynamics during the colonial period.

Industrialization and the Democratization of Dining Etiquette

The Industrial Revolution brought profound changes to social life, including dining practices. As urbanization expanded and the middle class grew, dining etiquette began to shift from exclusive aristocratic codes to more widely shared social norms. Public dining spaces such as restaurants became increasingly common, requiring standardized manners suitable for mixed social settings. Etiquette guides were published for broader audiences, emphasizing respectability, cleanliness, and polite interaction. Dining etiquette during this period reflected changing ideas about class, professionalism, and social mobility. It served as a means for individuals to navigate new social environments and assert their place within a rapidly transforming society.

Modern Developments and Cultural Continuity

In the modern era, dining etiquette continues to evolve while retaining historical influences. Globalization has introduced greater cultural diversity at the table, leading to more flexible and inclusive approaches to dining behavior. However, many traditional practices remain deeply valued, particularly in formal settings and ceremonial occasions. Dining etiquette today balances respect for tradition with adaptability to contemporary lifestyles. The persistence of certain customs, such as formal place settings or rituals of hospitality, demonstrates the enduring historical significance of dining etiquette as a cultural expression.

Napkin Folding as a Functional Design Practice

Napkin folding endures as a masterclass in functional design, distilling complex needs into elegant simplicity. From historical necessities to futuristic adaptations, it proves that profound utility hides in everyday gestures. By prioritizing human scale, material honesty, and contextual wit, it inspires broader design fields—from architecture's foldable facades to product's ergonomic interfaces. In an era of excess, reclaiming this practice reminds us that the most impactful designs are those we touch daily, quietly perfecting life's rituals.

Modern napkin folding embraces sustainability, favoring reusable fabrics over disposables, with folds engineered for longevity—reinforced corners withstand repeated creasing, extending lifecycle. Biodegradable papers employ wax-resist pleats for water resistance, minimizing waste in events. Digital aids like apps simulate folds for precise replication, democratizing expertise and reducing trial fabric errors.

Future innovations might integrate smart textiles with moisture-wicking folds that change color to indicate saturation, prompting timely changes hygienically. In zero-waste paradigms, modular folds disassemble into patterns for upcycling into scarves, closing material loops. This evolution positions napkin folding as vanguard functional design, responsive to ecological urgencies while honoring artisanal roots.

Historical Roots and Evolutionary Functionality

The history of napkin folding traces back to ancient civilizations, where it evolved from rudimentary wrappings to sophisticated structures driven by functional imperatives. In medieval Europe, large napkins known as "serviettes" were folded into elaborate shapes to contain spills during communal feasts, their bulk providing absorbent barriers against greasy foods common in those times. Renaissance banquets further refined this, with folds like the "bishop's hat" designed not just for visual drama but to create stable stands that kept napkins accessible and off damp tabletops, preventing sogginess and bacterial growth. By the 17th century, French court etiquette codified folds such as the "rose" or "water lily," which doubled as personal place markers, reducing mix-ups in crowded settings and streamlining service flow for harried staff. These historical adaptations highlight how napkin folding adapted to societal shifts—from feudal halls needing durable, reusable forms to Victorian parlors favoring compact, hygienic pockets that fit neatly into napkin rings for storage. Far from frivolous, this evolution mirrors broader design histories, akin to how flatware progressed from knives to specialized utensils, always prioritizing efficiency in human interaction with meals.

Principles of Functional Design Embodied in Folds

Functional Design in Napkin Folding: Adheres to principles like affordance, where the fold intuitively signals its use— a tall candle fold invites gripping at its base for wiping hands, while a shallow boat cradles utensils securely. Material science plays a pivotal role; cotton or linen's drape allows for self-supporting structures through precise pleating, harnessing gravity and friction to maintain shape without adhesives, embodying sustainable engineering.

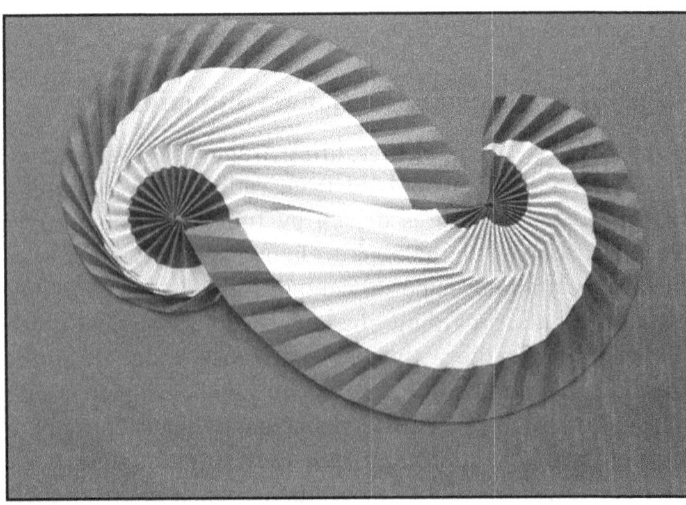

Fig: Intricacies of a Universal Design Element

Ergonomics Dictates Fold Dimensions: Edges must align with hand spans for comfortable grasping, and layers provide graduated absorbency, thicker at the center for heavy spills and thinner peripherally for precision dabs. Symmetry ensures balance, preventing topples that could disrupt dining harmony, while modularity allows disassembly into flat forms for laundering.

These elements converge in designs like the "envelope," which not only conceals menu cards but also unfolds into a full lap cloth, maximizing utility from limited fabric. In essence, each crease is a calculated decision, optimizing for durability, ease of execution, and adaptability across casual brunches to formal galas, proving that functionality thrives on constraint.

Techniques that Prioritize Utility over Ornament

Among the myriad techniques, the lotus fold stands out for its practical genius, beginning with a diagonal crease to form a pyramid base that elevates the napkin above condensation pools on plates, then layering petals that fan out to catch drips mid-meal. This design facilitates one-handed access, crucial for right-handed diners maneuvering forks, and collapses flat post-use, streamlining cleanup. Similarly, the goblet fan emerges from accordion pleats inserted into glass stems, not merely decorating stemware but absorbing condensation trickles that would otherwise slick tabletops, enhancing grip safety for wine glasses.

The pocket fold, quartered and notched, secures silverware vertically, reducing clatter and preventing rolls from contaminating surfaces, a boon in buffets where items shift during transport. Even complex bird-of-paradise variants prioritize airflow channels between wings to dry faster, extending usability in humid climates. Novices might start with the simple roll, twisted at one end to form a handle that prevents unraveling in laps, evolving to the swan where neck curvature guides wiping motions ergonomically. Each technique demands practice in tension control—over-tight folds crack under use, under-tight ones flop—refining the folder's skill into instinctive functionality craftsmanship.

Enhancing the Dining Experience through Practical Innovation

In contemporary table settings, napkin folding elevates functionality by orchestrating the entire meal workflow. A well-folded napkin at place settings signals readiness, psychologically priming guests for engagement while physically buffering against inevitable messes from soups or sauces. During multi-course meals, folds like the waterlily allow incremental unfolding, matching napkin expanse to spill scale—from pinky-tip pats for appetizers to full-lap coverage for mains—conserving fabric and reducing laundry volume.

In hospitality, these practices optimize labor; a stable knight's head fold holds multiple items, minimizing resets between seatings and cutting operational costs. For home cooks, the candle's cylindrical form doubles as a timer cue, its unrolling marking course transitions. Accessibility integrates seamlessly: larger folds for arthritic hands offer bigger grips, while color-coded pleats assist visually impaired diners in locating theirs. Beyond protection, folds influence sensory flow—crisp linen edges frame plates like picture borders, directing visual focus to cuisine, while subtle textures from folds add tactile pleasure, enriching the multisensory feast. This orchestration transforms dining from chore to choreography, where every tuck serves the human narrative.

Psychological and Social Dimensions of Functional Folds

Napkin folding's functionality extends to the psyche, fostering mindfulness and social cohesion through deliberate preparation. The act of folding instills calm focus, akin to meditative practices, reducing host stress via tangible progress markers. At the table, varied folds denote hierarchy subtly—taller structures for honored guests signal respect without words—easing conversational icebreakers and mitigating awkwardness in diverse gatherings. Psychologically, a thoughtfully folded napkin conveys care, boosting perceived meal quality and guest satisfaction, as studies in hospitality design correlate table aesthetics with tipping rates.

Socially, it democratizes etiquette; simple functional folds empower novices to host elegantly, bridging class divides historically enforced by ornate serviette artistry. In group dynamics, uniform folds synchronize the visual field, promoting unity and reducing decision fatigue over personal napkin placement. This layered impact underscores how functional design permeates intangible realms, turning cloth into a social lubricant.

Napkin Folding in Academic and Vocational Studies

Napkin folding in academic and vocational studies represents far more than a decorative exercise. It is a structured learning activity that embodies historical tradition, pedagogical value, cultural meaning, and professional relevance. By integrating napkin folding into curricula, educational institutions foster holistic skill development that balances intellectual understanding with practical competence. As hospitality and service industries evolve, the academic and vocational study of napkin folding remains a meaningful and enduring component of professional education.

Historical and Disciplinary Foundations

The academic relevance of napkin folding can be traced to its roots in European courtly traditions, where table presentation symbolized status, refinement, and order. Over time, these practices were codified into hospitality curricula and vocational training programs. In academic studies, napkin folding is often embedded within disciplines such as hotel management, home science, culinary arts, and event management. It provides a historical lens through which students explore evolving dining customs while also engaging with practical demonstrations that reinforce theoretical learning.

Napkin Folding as a Pedagogical Tool

In vocational education, napkin folding functions as an effective pedagogical tool that integrates cognitive, psychomotor, and affective learning domains. Students must understand instructions, visualize forms, and execute precise movements, thereby developing concentration and manual dexterity. Academically, instructors use napkin folding exercises to teach sequencing, symmetry, and design logic.

The repetitive yet creative nature of the practice helps learners internalize concepts of order, patience, and attention to detail, which are transferable to other professional skills. Fabric napkins add a touch of elegance to your holiday meals. For most of us, it's only during these meals that we even bother with fabric napkins. Instead of folding them in a rectangle or square this year, impress your guest with some fancy napkin folding skills. Don't worry — these three folds are easy to do and will make a dinner guest feel like they are dining at a 5-star restaurant. For best folding results, it will help to first iron the napkins thoroughly or even lightly starch the fabric to give the napkin a more rigid form. Linen and cotton napkins are the easiest to fold.

The Cone

Fig: Simple Fold is Elegant and Easy to do

Follow the numbered steps below with images ordered from left:

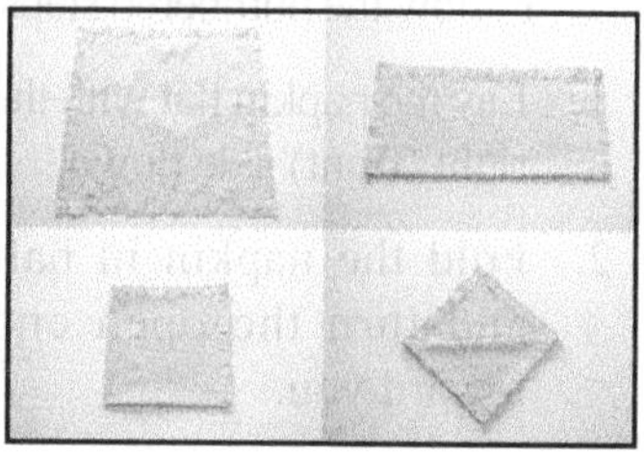

1. Lay the napkin flat with the right (front) side down.
2. Fold the napkin in half and turn the open end away from you.
3. Fold the napkin into quarters. Turn the napkin so that the open end is facing away from you.
4. Fold the top layer all the way forward and press it down flat.

5. Turn the napkin over and fold the left-hand side of the napkin over two-thirds of the way (image at left below).

6. Then fold the right-hand side of the napkin over to match the left-hand side. Tuck the corners of the napkin into the left-hand side flap. Flip it over and there you have it — the cone. Tuck an herb or flower sprig in the pocket or flatware.

The Flatware Pocket

Fig: Create a Little Pocket to Tuck Flatware for this Classic Presentation

Follow the numbered steps below with images ordered from left:

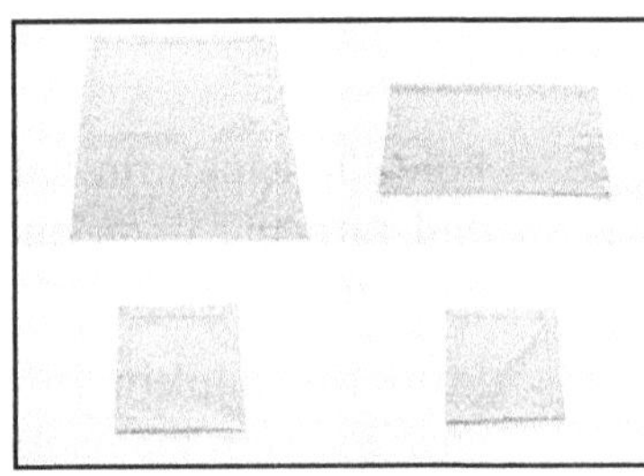

1. Lay the napkin flat with the right (front) side down.
2. Fold the napkin in half and turn the open end toward you.
3. Fold the napkin into quarters. Turn the napkin so the open corner is facing away and to the left.
4. Fold the top layer of napkin in half diagonally and press it down.

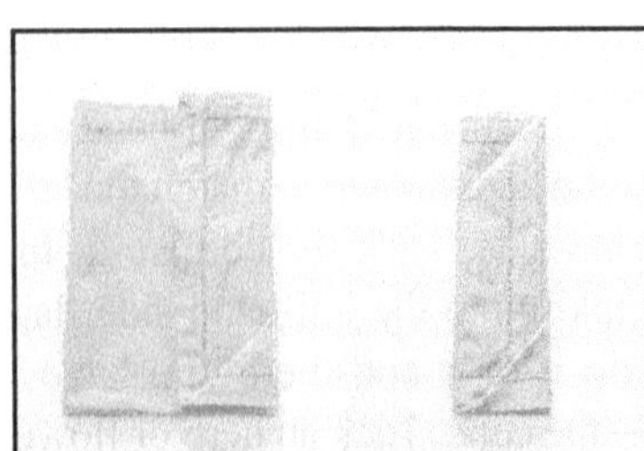

5. Fold the right side back one-third of the way and press it down (image at left below).
6. Then, fold the left side back also one-third of the way and press. Flip the napkin over and insert your flatware.

Three Point Fold

Fig: Napkin Fold Looks Pretty on its Own, But you can Embellish it with Greenery or Even a Piece of Fruit Laid on Top of the Napkin

Follow the numbered steps below with images ordered from left:

1. Lay the napkin flat with the right (front) side down.
2. Fold the napkin in half diagonally with the middle point of the triangle facing down.

Fold the right corner of the napkin over to the left.

3. Then, fold the right side of the napkin towards the left until the desired "necktie shape" is formed on the napkin.

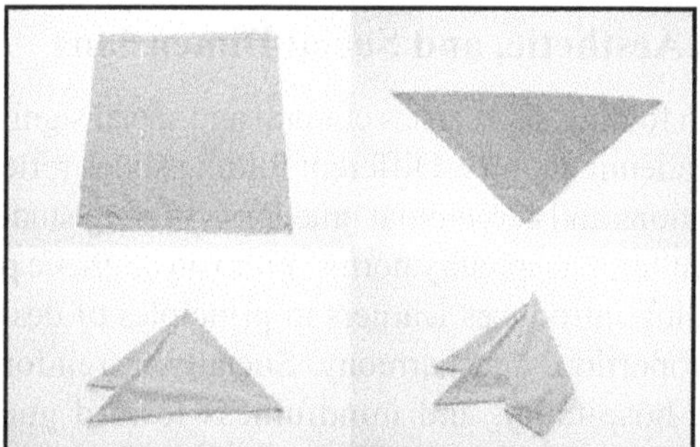

Fig: Any of these Folds will Add a Touch of Elegance to your Table

Relevance in Hospitality and Service Education

Hospitality education places strong emphasis on service quality and guest perception, both of which are influenced by table presentation. Napkin folding is therefore treated as a core competency in vocational hospitality programs. Through structured practice, students learn how folded napkins contribute to ambiance, brand identity, and customer satisfaction. Academic discussions often link napkin folding to service psychology, explaining how visual cues affect diners' expectations and experiences. This dual focus strengthens both conceptual understanding and workplace readiness.

Skill Development and Employability Outcomes

One of the primary reasons napkin folding remains relevant in vocational studies is its contribution to employability. Employers in hospitality, catering, and event industries expect graduates to demonstrate refined presentation skills. Napkin folding allows students to showcase professionalism, creativity, and adherence to standards during practical assessments. Academically, it also supports outcome-based education by providing measurable performance indicators, such as precision, consistency, and aesthetic appeal, which can be evaluated objectively.

Cultural, Aesthetic, and Social Dimensions

Napkin folding also carries cultural and social significance that enriches academic inquiry. Different folding styles reflect regional dining traditions and ceremonial practices, offering students insight into cross-cultural hospitality norms. From an aesthetic perspective, napkin folding introduces learners to principles of design such as balance, proportion, and harmony. Socially, it reinforces values of respect, hospitality, and mindfulness toward guests. These dimensions make napkin folding a multidisciplinary topic that extends beyond technical training.

Integration into Curriculum and Assessment

Educational institutions integrate napkin folding into curricula through demonstrations, workshops, and practical examinations. In academic settings, it may form part of modules on food service operations, etiquette studies, or applied design. Vocational assessments often evaluate students on technique, speed, and presentation consistency. This integration ensures that learners not only acquire the skill but also understand its theoretical justification and professional relevance within broader course objectives.

Table: Comparative Perspectives in Academic and Vocational Studies

Aspect	Academic Studies	Vocational Studies
Primary Focus	Conceptual understanding and cultural context	Practical skill mastery and workplace readiness
Teaching Approach	Lectures, discussions, and demonstrations	Hands-on training and repetition
Assessment Method	Written exams and reflective analysis	Practical tests and performance evaluation
Learning Outcome	Analytical and theoretical insight	Job-oriented competence
Application	Research, teaching, and design studies	Hospitality, catering, and event services

Learning Outcomes Associated with Napkin Folding

The structured teaching of napkin folding enables learners to achieve several important outcomes, which are commonly emphasized across academic and vocational programs:

- Development of fine motor skills and hand-eye coordination.
- Enhanced understanding of dining etiquette and service standards.

- Improved aesthetic judgment and design awareness.
- Strengthening of patience, focus, and discipline.
- Practical exposure to hospitality industry expectations.
- Application of theoretical concepts to real-world scenarios.
- Increased cultural awareness through diverse folding styles.
- Reinforcement of professional grooming and presentation values.
- Contribution to teamwork during service simulations.
- Boost in learner confidence during practical evaluations.

Contemporary Relevance and Educational Value

In modern education, where experiential learning is increasingly prioritized, napkin folding remains a valuable instructional activity. It aligns with competency-based education models that emphasize demonstrable skills alongside academic knowledge. Technological advancements and changing dining trends have not diminished its importance; rather, they have expanded creative possibilities and design interpretations. For vocational learners, napkin folding continues to be a visible marker of service excellence, while for academic students, it offers a tangible connection between theory, culture, and practice.

Interdisciplinary Perspectives on Table Presentation

Aesthetics of Arrangement and Visual Rhythm

From the perspective of art history and design theory, table presentation is an exercise in composition and balance. The table surface functions as a canvas, governed by the same principles of the golden ratio, symmetry, and negative space that guide a painter. A well-presented table avoids the chaos of clutter and the sterility of emptiness, aiming instead for a state of "controlled abundance." This requires an understanding of visual weight; a heavy, dark entree on the left must be balanced by the visual mass of the bread plate or the sheen of the wine glass on the right. The designer must consider the verticality of the arrangement as much as the horizontal.

Fig: Folding Dinner Napkin

The height of a candleholder or the arch of a stemware glass introduces rhythm, creating a skyline that guides the eye across the setting. This is gastrophysics in its purest form—the application of artistic principles to the sensory experience of eating. The placement of a garnish is not random; it is a focal point, a period at the end of a sentence that provides closure to the visual composition of the dish. Without this artistic scaffolding, the meal risks becoming a disjointed collection of ingredients rather than a unified aesthetic object.

Psychology of Perception and Sensory Transference

Perhaps the most fascinating interdisciplinary angle is that of cognitive psychology and neuroscience, specifically the study of how vision hijacks taste. Research in this field consistently proves that we eat with our eyes first, and table presentation is the primary tool for manipulating this expectation. The shape of the plate, for instance, alters our perception of portion size and flavor intensity.

Food served on a round plate is perceived as sweeter than the exact same food served on a square plate with sharp angles, a phenomenon known as sensory transference. The color of the crockery interacts with the food to create contrast or harmony; a white plate makes colorful vegetables appear more vibrant, triggering a subconscious association with freshness and health. Furthermore, the weight of the cutlery creates a haptic feedback loop; heavy silverware implies expensive, high-quality food, actually causing the brain to release more dopamine and making the food taste better. Table presentation is, therefore, a form of pre-digestive manipulation, priming the brain for a specific culinary experience before the tongue is even involved.

Sociological Signaling and the Language of Status

Sociology offers a harsher, more critical lens through which to view the set table, interpreting it as a map of social hierarchy and exclusion. The arrangement of the table is rarely democratic; it is a spatial representation of power. The head of the table, usually reserved for the host or the person of highest status, is physically elevated or centrally located, commanding the room. The distance from this center point dictates one's importance.

The number of forks and glasses is not just about utility; it is a display of cultural capital. Knowing which fork to use for the fish course is a shibboleth, a secret password that signals membership in the upper echelons of society. Historically, the separation of the sexes and the strict ordering of service were methods of social control. Even today, the "mise-en-place" acts as a gatekeeper. A table set with esoteric implements can intimidate the uninitiated, reinforcing class barriers. Conversely, the rejection of formal presentation—the family-style platter—signals intimacy and egalitarianism. Thus, the table setting is a sociological script that tells us exactly where we stand in relation to one another.

Historical Palimpsests and the Archaeology of Ritual

Every knife, fork, and napkin fold are a historical artifact, a palimpsest where centuries of dining rituals are written over one another. To understand table presentation, one must be an amateur historian. We still speak the language of medieval banquets when we use the "trencher"—that is, the dinner plate—which was originally a slice of stale bread used as a disposable plate. We retain the Victorian obsession with silverware, even in an age of casual dining.

The interdisciplinary study of history reveals that our current obsession with "plating" is actually a relatively new phenomenon. For most of human history, food was piled high in the center of the table, and the individual plate did not exist. The shift to the individual plate in the 17th century coincided with the rise of the nation-state and the individual self. The fork, once considered an effeminate and blasphemous tool for interfering with God's food, became a symbol of civilization. When we set a table today, we are unconsciously reenacting the power struggles of Louis XIV's court. The presentation is a museum of dead customs, preserved and repurposed for modern consumption.

Material Science and the Haptic Experience

Finally, we must look to material science and physics to understand the "feel" of the table. The dining experience is profoundly haptic, relying on the interaction between the body and the object. The thermal conductivity of the material matters immensely; a ceramic plate keeps food warm, while a metal tray cools it rapidly, changing the texture of fats and oils. The acoustic properties of the table are equally vital. The "clink" of crystal versus the dull thud of glass creates a different emotional resonance. High-frequency sounds are associated with luxury and crispness, while low frequencies suggest heaviness. The friction coefficient of the tablecloth determines how easily the plate slides, affecting the diner's sense of stability. Material scientists study how the micro-texture of a glaze affects the release of aromas. A rough, unglazed stoneware might trap aromas, while a smooth, vitrified porcelain allows them to waft up to the nose. Table presentation is not just visual; it is a tactile and acoustic architecture that engages the body's somatosensory system, proving that the vessel is just as important as the contents.

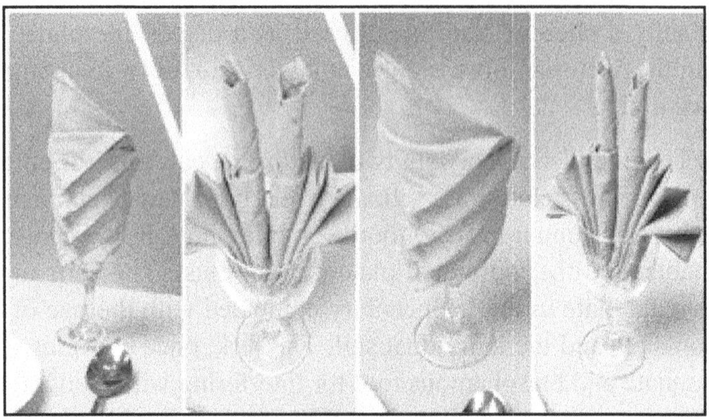

Fig: Material Science in the Tactile Experience of Napkin Folding

2

Historical Evolution of Napkin Folding Practices

Origins of Table Linen in Ancient Civilizations

The origins of table linen in ancient civilizations reveal the deep connections between material culture, social values, and everyday practices. From linen in Egypt to silk in China and cotton in the Indus Valley, textiles played a vital role in shaping dining environments and expressing cultural ideals.

Table linen served practical, symbolic, and social functions, reflecting concerns with cleanliness, status, and hospitality. By examining these early uses, one gains a richer understanding of how ancient societies transformed simple cloth into a meaningful element of communal life, leaving a legacy that continues to influence dining traditions across the world.

Early Textile Production and its Social Significance

The emergence of table linen was made possible by early advances in textile production. Ancient societies such as those in Mesopotamia, Egypt, and the Indus Valley mastered the spinning and weaving of natural fibers like flax, cotton, and wool. Linen, derived from flax, became particularly significant due to its durability, lightness, and association with cleanliness. Textile production was often labor-intensive and time-consuming, making finely woven cloth a symbol of wealth and refinement. The availability of textiles influenced domestic practices, including how food was served and consumed, leading to the gradual incorporation of cloth into dining spaces.

Table Linen in Ancient Egypt

Ancient Egypt offers some of the earliest evidence of the use of linen in domestic and ceremonial contexts. Linen held a special place in Egyptian culture, associated with purity, light, and divine favor. It was widely used in clothing, religious rituals, and burial practices, and its use extended naturally to dining settings.

Wealthy households used linen cloths to cover tables or serving surfaces, enhancing both cleanliness and aesthetic appeal. The presence of table linen in banquets and feasts symbolized social status and respect for guests. In religious contexts, linen-covered surfaces were used in offerings to gods, reinforcing the idea that linen represented order and sacredness in everyday life.

Mesopotamian Practices and the Role of Cloth in Dining

In Mesopotamia, one of the world's earliest urban civilizations, communal meals were central to social and religious life.

Fig: According to other Mesopotamian Cuneiform Tablets

Archaeological evidence suggests that textiles played a role in these gatherings, particularly in elite households and temples. Cloths were used to cover low tables or platforms on which food was placed, protecting surfaces and signifying care in presentation. In a society that valued ritual order and hierarchy, the use of textiles during meals reflected the importance of structure and propriety. Table linen in Mesopotamia was closely linked to temple economies, where textile production and distribution were controlled by religious institutions, highlighting the connection between cloth, food, and sacred authority.

Indus Valley Civilization and Early Cotton Use

The Indus Valley Civilization is notable for its early and sophisticated use of cotton textiles. Unlike many other ancient societies that relied primarily on flax or wool, the people of the Indus Valley developed advanced techniques for cotton cultivation and weaving.

Although direct evidence of table linen is limited due to the perishable nature of textiles, depictions and artifacts suggest that cloth played an important role in domestic life. The use of cotton cloth in dining spaces likely reflected an emphasis on cleanliness and order within the household. The presence of standardized urban planning and sanitation systems in Indus cities supports the idea that textile use, including table coverings, was part of a broader cultural commitment to hygiene and organization.

Ancient China and Ritualized Dining Practices

In ancient China, dining was deeply ritualized and closely connected to philosophical and social principles. Textiles were an integral part of these rituals, particularly in elite and ceremonial settings. Silk, the most prized Chinese textile, was often used to cover tables and offering surfaces during important feasts and ancestral rituals. The use of fine cloths in dining contexts reflected Confucian values of respect, hierarchy, and harmony. Table linen helped define spaces of honor and reverence, distinguishing ordinary meals from those with ritual significance. The careful placement and maintenance of textiles during meals demonstrated moral discipline and social awareness.

Greek and Roman Contributions to Table Linen Traditions

The civilizations of ancient Greece and Rome played a significant role in shaping later Western dining customs, including the use of table linen. In ancient Greece, meals were often taken on low tables, and cloths were used primarily for wiping hands and surfaces rather than fully covering tables. These early napkin-like textiles represented a growing concern for cleanliness and personal comfort.

In Roman society, the use of tablecloths became more widespread, particularly during formal banquets. The Roman mensa was often covered with a cloth, both to protect the table and to enhance the visual appeal of the meal. Roman table linen also served a practical purpose, as guests used sections of the cloth or separate napkins to wrap leftover food, highlighting the multifunctional role of textiles in dining.

Fig: Ancient Greek Feast with Actors and Musicians

Symbolism and Status Associated with Table Linen

Across ancient civilizations, table linen carried symbolic meanings that extended beyond practicality. The quality, color, and material of cloth used in dining contexts often indicated social rank and wealth. Fine linen or silk was associated with elites, while coarser textiles were used in more modest households. Table linen also reflected values of hospitality, as providing clean and well-maintained cloths demonstrated respect for guests. In many cultures, the presence of linen at the table marked the transition from everyday eating to a more formal or ceremonial occasion, reinforcing social bonds and cultural identity.

Continuity and Influence on Later Traditions

The use of table linen in ancient civilizations laid the foundation for later dining traditions in medieval and modern societies. Techniques of textile production, symbolic associations, and practical applications were passed down and adapted over time. As trade expanded and cultures interacted, ideas about table linen spread across regions, influencing new forms and uses. The enduring presence of tablecloths and napkins in contemporary dining reflects the lasting legacy of ancient practices, demonstrating how early innovations in textile use continue to shape social rituals today.

Medieval Dining Customs and Early Folding Forms

Medieval dining customs forged early napkin folding forms as bastions of order amid chaos, their utility etching enduring design paradigms. From hygienic bulwarks to status sigils, these practices humanized feasts, paving civility's path. Their legacy endures, whispering of eras when every crease conquered crudity.

Communal Feast and Hygienic Imperatives

Medieval dining unfolded in cavernous great halls where

dawn-to-dusk labors converged in evening repasts, customs dictating that lords and retainers shared trenchers passed hand-to-hand, fingers serving as utensils. Without forks until the late period, greasy haunches and pottages necessitated robust napkins draped across laps or suspended from belts, their folds creating pouches to capture errant broths that sloshed from wooden bowls.

Hygiene was paramount in plague-shadowed times; folds elevated fabric from filthy floors, preventing recontamination when retrieved, while pleated edges wicked moisture away from skin. Monks in cloistered silence adhered to stricter codes, folding napkins into compact rectangles post-meal for ritual stacking, a practice echoing military precision. These customs stemmed from Carolingian edicts mandating clean linens, transforming napkins from luxury imports to ubiquitous tools, their early folds—simple diagonals or rolls—born of pragmatism to contain vast cloths against wind-whipped drafts.

Serviettes in Feudal Hierarchy and Table Layout

Feudal dining reinforced social strata through napkin deployment, where size and placement bespoke rank. A baron's serviette might span arm's length, folded lengthwise into a bolster to shield embroidered tunics from falconry-tainted fingers, while vassals received narrower strips rolled into wands for belt tucks.

Trestle tables, assembled for feasts and dismantled post-prandially, featured napkins pinned at ends to form tents over bread trenchers, shielding against flies and soot from central hearths. In tournaments or weddings, excess demanded innovation; multiple napkins layered in staggered folds absorbed progressive soiling from fish courses to sweetmeats. Heraldic embroidery guided folding axes, aligning crests upward for visibility, merging heraldry with utility. This hierarchy infused folds with intent, ensuring even the lowborn glimpsed civility's promise through borrowed grandeur.

Birth of Structured Folding Techniques

Early folding forms crystallized amid 12th-century Norman influences, where crusaders imported Levantine wrapping arts, adapting them to heavy linens. The foundational "sallet" fold quartered the square into a pouch mimicking a helmet, sturdy enough to hold salt cellars aloft on uneven boards, preventing spills into communal stews.

Diagonally creased "bishop's mitre" rose pyramidally, its peak signaling clergy precedence while base layers sopped wine sloshes from horn cups. Roll-and-twist methods dominated daily use, coiling napkins into sausages tucked under arms for easy access during hunt banquets, unraveling sequentially as hands fouled. These techniques exploited linen's weave for self-support, tensioned pleats mimicking basketry to cradle fruits or nuts between courses. Manuscripts like the 14th-century Forme of Cury hint at such folds, prescribing them alongside recipes for synchronized presentation.

Fig: Art and Engineering Fold into Inspiration

Courtly Refinements and Etiquette Manuals

By the late medieval courts of Burgundy and England, etiquette tracts elevated folding to protocol, with folds like the “water lily” budding from radial creases to fan protectively over finger bowls introduced circa 1300. These manuals, penned by stewards, detailed sequences: start with starched cloth star-folded for crispness, then sculpt into boats ferrying conserves across tables slick with mead.

Women’s smaller “essuie-manche” folded into cuffs shielded billowing sleeves during spiced wine toasts, embodying gendered pragmatism. Chivalric orders mandated identical folds for knights’ tables, fostering brotherhood through uniformity, while troubadour tales romanticized the deft folder as courtly virtuoso. Such refinements tempered barbarism, folds acting as tactile mnemonics for behavioral codes amid raucous minstrelsy.

Functional Roles across Meal Progressions

Fig: Napkin Folds for your Table

Napkin folds adapted dynamically through meals’ arc, from appetizers of salt-fish in coiled nests that drained brine to main platters where bolster folds buttressed elbows against jostling. Dessert phases unfurled fans to dust sugar from beards, their expansive wings catching crumbs from gingerbread. In winter castles, insulated folds retained warmth from hearth proximity, doubling as mufflers. Travel feasts on campaign trails compressed napkins into quiver-like sheaths for jerky wipes, portability trumping permanence. Illness-struck tables saw medicinal folds pouched with possets, hygiene paramount. This versatility underscores folds' role as meal choreographers, pacing indulgence with practicality.

Symbolism, Superstition, and Cultural Interplay

Folds carried layered meanings, triangular forms invoking the Trinity for pious feasts, warding ill omens from asymmetric tables. Superstitions dictated odd creases to deflect envy at opulent spreads, while intercultural exchanges—via Silk Road silks or Moorish serviettes—infused Moorish arabesque pleats, softening angular Christian geometries. Jewish medieval tables folded into menorah silhouettes for Sabbath, blending faith with function. These symbols wove folds into cultural tapestries, transcending mere cloth.

Legacy into Renaissance Transformations

As medieval waned, early folds seeded Renaissance opulence, Burgundian splendor birthing filigreed variants from crude sallets. Printing presses disseminated patterns, perpetuating customs into printing-era homes. This bridge cemented napkin folding's trajectory from feudal fix to fine art.

Renaissance Influence on Decorative Table Settings

Renaissance influence on decorative table settings represents a significant moment in the history of dining culture, where art, philosophy, and social life intersected at the table. By applying humanist ideals, artistic principles, and refined craftsmanship to dining spaces, Renaissance society elevated the act of eating into an expression of cultural identity and intellectual aspiration. The enduring impact of these practices demonstrates how decorative table settings can serve as powerful reflections of broader cultural transformations, linking the past to present traditions of elegance and hospitality.

Cultural and Historical Context of the Renaissance

The Renaissance, spanning roughly from the fourteenth to the seventeenth century, was characterized by a revival of classical Greco-Roman ideas and a renewed focus on human potential and beauty. This cultural shift influenced everyday practices, including dining customs. Banquets became carefully orchestrated displays of power and learning, particularly in Italian city-states such as Florence and Venice. Decorative table settings were designed to echo the symmetry, proportion, and harmony admired in Renaissance art, reinforcing the idea that beauty and order should permeate all aspects of life.

Humanism and the Art of the Table

Humanism, a defining philosophy of the Renaissance, emphasized the dignity, intellect, and sensory experiences of human beings. This worldview transformed table settings into expressions of cultivated taste and intellectual awareness. Decorative objects such as finely embroidered tablecloths, sculpted salt cellars, and engraved utensils reflected both artistic skill and scholarly inspiration. Classical motifs, mythological figures, and geometric patterns were incorporated into tableware, symbolizing a connection between daily life and the intellectual ideals of the era.

Materials, Craftsmanship, and Artistic Detail

Renaissance decorative table settings showcased exceptional craftsmanship and the use of luxurious materials. Linen tablecloths were prized for their whiteness and intricate embroidery, often featuring floral or heraldic designs. Tableware made from silver, pewter, and occasionally gold demonstrated both wealth and artistic patronage. The emphasis on craftsmanship mirrored the Renaissance belief in the value of skilled labor and artistic mastery. Each element placed on the table was intended to contribute to a cohesive visual narrative of refinement and order.

Symbolism and Social Hierarchy at the Table

During the Renaissance, table settings also functioned as symbolic representations of social rank and authority. The placement of objects, seating arrangements, and decorative elements communicated status and power. Hosts used elaborate table decorations to assert their cultural sophistication and political influence. The symmetry and balance characteristic of Renaissance aesthetics reinforced hierarchical order, reminding guests of their position within the social structure while simultaneously impressing them with visual splendor.

Influence of Art and Architecture on Table Design

Renaissance art and architecture had a direct impact on the design principles applied to table settings. Concepts such as perspective, proportion, and symmetry guided the arrangement of table elements. Centerpieces were designed to draw the eye, much like focal points in paintings or architectural spaces. Decorative objects often mirrored architectural forms, such as columns or arches, translating monumental artistic ideas into miniature, functional designs suitable for the dining table.

Role of Banquets and Ceremonial Dining

Fig: Large Ceremonial Banquet Hall

Banquets during the Renaissance were elaborate events that blended gastronomy, performance, and visual art. Decorative table settings played a central role in these ceremonies, enhancing the overall sensory experience. Sculptural centerpieces made from sugar, bread, or wax were commonly used to demonstrate creativity and abundance. These temporary artworks reflected the Renaissance fascination with spectacle and transformation, reinforcing the idea that dining was both a cultural and artistic performance.

Table: Pre- and Renaissance Table Settings

Aspect	Medieval Table Settings	Renaissance Table Settings
Aesthetic Focus	Functionality and symbolism	Beauty, balance, and harmony
Materials Used	Wood, simple metals	Linen, silver, artistic metals
Decorative Style	Minimal and utilitarian	Elaborate and artistic
Cultural Influence	Religious tradition	Classical and humanist ideals
Social Expression	Basic hierarchy	Refined display of status

Characteristics of Renaissance-Inspired Table Settings

The influence of the Renaissance on decorative table settings can be summarized through several defining characteristics that continue to inspire modern dining aesthetics:

- Emphasis on symmetry and balanced arrangement.
- Use of fine linens with detailed embroidery.
- Incorporation of classical and mythological motifs.

- Display of artistic craftsmanship in tableware.
- Symbolic use of objects to convey status.
- Integration of art and functionality.
- Focus on visual harmony and proportion.
- Celebration of abundance and hospitality.
- Reflection of humanist values through design.
- Transformation of dining into a cultural performance.

Legacy and Contemporary Relevance

The Renaissance influence on decorative table settings has endured far beyond its historical period. Modern formal dining practices continue to draw inspiration from Renaissance principles of balance, elegance, and artistic expression.

Contemporary table design often echoes Renaissance ideals through coordinated color schemes, carefully chosen centerpieces, and an emphasis on presentation. In academic studies, this legacy provides insight into how cultural movements shape everyday practices, while in design and hospitality education, it offers a foundation for understanding the relationship between art, culture, and social interaction.

Napkin Folding in European Aristocratic Traditions

The folded napkin is a time capsule, preserving the values, anxieties, and aspirations of the European aristocracy. It tells the story of a class that sought to impose order on chaos, to elevate the mundane to the magnificent. Each fold, each crease, was a testament to their belief in their own superiority, a belief written in linen and starch. Today, when we encounter a meticulously folded napkin, we are not just seeing a piece of tableware; we are touching history. It is a silent witness to the rise and fall of empires, a humble cloth that once held the weight of the world.

Origins in the Medieval Hall: From Communal Cloth to Personal Emblem

The story of the aristocratic napkin begins in the drafty halls of the Middle Ages, where dining was a communal, chaotic affair. Here, the "napkin" was a shared cloth, often a rough piece of linen passed around the table for wiping hands and blades. It was the Renaissance that introduced the concept of the individual napkin, a shift mirrored in the rise of humanism and the individual as a distinct entity.

By the 16th century, Italian courts, influenced by the rediscovery of classical Roman dining practices, began to fold napkins into simple shapes—a triangle, a roll—as a mark of personal space and hygiene. This was the first whisper of the napkin as a status symbol. When Catherine de' Medici brought her Italian habits to the French court, she inadvertently planted the seed of what would become a full-blown obsession with napkin folding.

Sun King's Starch: Versailles and the Architecture of Linen

It was under Louis XIV that the napkin ascended to its zenith as a tool of aristocratic expression. At Versailles, the table was a stage, and the napkin its most versatile prop. The grand couvert, the ritualized meal where the king dined in public, demanded spectacle. Napkins were starched to a rigidity that bordered on the absurd, folded into towering structures that defied gravity and practicality.

The pliage a la royale—a fold so tall it resembled a spire—was reserved for the monarch, a literal elevation above his peers. Lesser nobles received less lofty folds, their napkins' height and complexity calibrated to their rank. This was not mere decoration; it was spatial politics. The napkin's architecture mirrored the court's hierarchy, its crisp edges and sharp angles a metaphor for the unyielding order of absolutism. To unfold such a napkin was unthinkable; it was meant to be admired, not used.

18th Century: Folds as Social Cipher

By the 18th century, napkin folding had become a language of its own, a semiotic system decipherable only by the initiated. Each fold carried a specific meaning, a code understood by the aristocracy but opaque to outsiders. The "Bishop's Mitre," for instance, signaled the presence of a high-ranking cleric, while the "Crown" fold was reserved for royalty.

The "Lotus" fold, with its delicate petals, might denote a celebration, while a simple, flat fold suggested mourning. This was not whimsy; it was a form of nonverbal communication, a way to convey messages without uttering a word. The maître d'hôtel, the architect of these folds, was a figure of immense skill and discretion. He knew not only how to manipulate linen but also how to read the social currents of the table, translating them into the silent poetry of the napkin.

Fig: Unfolding the History of Napkin Art

Victorian Apotheosis: Complexity as Virtue

The 19th century, with its Victorian obsession with order and display, saw napkin folding reach its apotheosis. This was the era of the haute bourgeoisie, the newly wealthy striving to mimic the aristocracy they sought to join. Napkins were folded into shapes of breathtaking complexity: swans, fans, candles, even entire floral arrangements. The "Standing Fan" fold, with its radiating pleats, became a staple of grand dinners, a symbol of hospitality and refinement. This was not merely about aesthetics; it was about labor. A napkin folded into a swan required time, skill, and patience—resources the aristocracy had in abundance. It was a way to flaunt one's ability to waste effort on the ephemeral, to transform a humble piece of linen into a work of art that would be destroyed with the first bite of soup.

Invisible Hand: Servants as Linen Architects

Behind every folded napkin stood a servant, whose labor was as invisible as it was essential. The valet, the lady's maid, the footman—these were the unsung artists of the aristocratic table. They knew the secrets of starch, the precise angle of the iron, the exact moment to dampen the cloth for a sharper crease. Their hands moved with the precision of surgeons, yet their work was never acknowledged.

The napkin's perfection depended on their skill, yet its purpose was to erase their presence. This invisibility was deliberate. The aristocracy's power rested on the illusion of effortlessness. A perfectly folded napkin was meant to appear as though it had folded itself, a natural extension of the table's elegance. The servant's role was to vanish into the background, leaving only the pristine linen as evidence of their toil.

Decline and Legacy: The Folded Napkin in a Changing World

The First World War marked the beginning of the end for the aristocratic napkin tradition. The collapse of monarchies, the rise of the middle class, and the disappearance of domestic servants rendered the elaborate folds obsolete. The 20th century demanded practicality, not pageantry. Yet the legacy of the folded napkin endures. The "Pyramid" fold still graces hotel tables, a ghost of Versailles. The "Standing Fan" appears at weddings, a nostalgic nod to Victorian grandeur. These folds are no longer symbols of power but relics of a bygone era, reminders of a time when even the way one wiped one's mouth was a matter of status.

Fig: Napkin Etiquette of Paper

Colonial and Global Diffusion of Folding Techniques

The colonial and global diffusion of folding techniques reveals how seemingly simple practices can carry deep cultural significance and historical complexity. Through trade, education, and domestic life, folding techniques traveled across borders, adapting to new environments and contributing to shared traditions. The interplay of exchange, adaptation, and resistance underscores the role of folding as both a practical skill and a cultural expression. Understanding this history enriches appreciation of folding techniques not merely as decorative arts but as living traditions shaped by centuries of global interaction and cultural dialogue.

Pre-Colonial Folding Traditions and Indigenous Knowledge

Before the onset of European colonial expansion, many societies possessed well-developed folding traditions rooted in indigenous knowledge systems. In East Asia, particularly in Japan and China, folding practices were closely linked to ritual, aesthetics, and craftsmanship. Textile folding in clothing storage, gift wrapping, and ceremonial presentation reflected philosophical ideals of balance and respect. In parts of Africa and the Middle East, cloth folding was embedded in domestic routines and hospitality customs, shaping how garments, table coverings, and ceremonial textiles were prepared and displayed. These practices evolved independently, responding to local materials, social structures, and cultural values, forming rich traditions long before global contact intensified.

European Colonial Expansion and Cultural Exchange

The age of European colonial expansion created unprecedented opportunities for cross-cultural exchange, including the transmission of folding techniques. As European powers established colonies in Asia, Africa, and the Americas, they encountered diverse material cultures that influenced their own domestic and ceremonial practices. Colonial administrators, missionaries, and traders observed and sometimes adopted local folding methods, particularly those related to textiles and presentation. At the same time, European customs were introduced into colonial societies, often through formal education, religious institutions, and elite households. Folding techniques thus became part of a broader exchange of cultural practices shaped by power relations and social hierarchies.

Maritime Trade Routes and the Spread of Techniques

Global maritime trade played a crucial role in the diffusion of folding techniques during the colonial period. Ships carried not only goods but also knowledge, skills, and aesthetic preferences. Sailors, merchants, and artisans transported textiles, manuals, and decorative objects that demonstrated specific folding styles. Ports became cultural crossroads where techniques from different regions converged and blended. In these settings, folding practices were often adapted to suit new materials and social contexts, leading to hybrid forms that combined local traditions with foreign influences. The movement of folding techniques along trade routes highlights the interconnectedness of material culture in the early modern world.

Colonial Households and the Formalization of Folding Practices

Fig: Colonial Homes and the Emergence of Formal Folding Traditions

Within colonial households, particularly those of European elites, folding techniques were formalized as markers of refinement and order. Table linen folding, in particular, gained prominence as a symbol of civility and domestic discipline. Colonial manuals on household management often included instructions on proper folding and presentation, reflecting European ideals of cleanliness and hierarchy. Indigenous servants and artisans were trained in these methods, leading to the incorporation of European folding styles into local domestic routines. Over time, these practices were localized, blending with existing traditions and contributing to distinctive regional expressions of folding.

Missionary Influence and Educational Transmission

Missionary institutions played a significant role in spreading folding techniques across colonial territories. Schools and training centers established by missionaries emphasized discipline, order, and moral instruction, often using domestic skills as tools for cultural assimilation.

Folding practices were taught as part of household training, particularly to women and girls, reinforcing gendered expectations and social roles. Through education, folding techniques were standardized and disseminated widely, extending their reach beyond elite households into broader segments of colonial society. This process illustrates how everyday practices became instruments of cultural transmission and social control.

Adaptation and Resistance in Colonial Contexts

While colonial diffusion introduced new folding techniques, local communities did not adopt them passively. Indigenous societies often adapted foreign practices to align with their own values and needs, resulting in creative reinterpretations. In some cases, traditional folding methods were preserved as acts of cultural resistance, maintaining continuity with pre-colonial identities. The coexistence of imported and indigenous folding techniques reflects the complex dynamics of colonial encounters, where cultural exchange was shaped by negotiation, adaptation, and selective acceptance rather than simple imitation.

Post-Colonial Globalization and Continued Diffusion

The end of formal colonial rule did not halt the global diffusion of folding techniques. Instead, post-colonial globalization accelerated cultural exchange through mass media, international travel, and global industries.

Folding practices associated with hospitality, dining, and ceremonial presentation became standardized in international settings such as hotels and restaurants, drawing on diverse cultural influences. At the same time, renewed interest in traditional crafts and cultural heritage led to the revival and global appreciation of indigenous folding techniques. These developments highlight the ongoing evolution of folding practices in a world shaped by both historical legacies and contemporary connections.

Twentieth-Century Innovations in Table Presentation

The twentieth century redefined table presentation through innovation, cultural exchange, and evolving values. From the rejection of Victorian excess to the embrace of minimalism, from material revolutions to sustainable practices, each development reflected broader societal shifts. Table presentation evolved from a rigid formality into a flexible, expressive art form that enhances the dining experience on visual, tactile, and emotional levels. As we enter the twentyfirst century, these innovations continue to shape how we set our tables—proving that even the simplest meal can be elevated by thoughtful, creative presentation.

Early Twentieth Century: Rejection of Excess and Embrace of Modernism

The dawn of the twentieth century marked a decisive break from the ornate, heavily ornamented tables of the Victorian and Edwardian eras. Influenced by the Arts and Crafts Movement and later by Art Nouveau and Art Deco, designers sought simplicity, honesty of materials, and functional beauty. The Arts and Crafts Movement, championed by figures such as William Morris, emphasized handcrafted, natural materials—woven placemats, hand-thrown ceramics, and linen tablecloths—rejecting industrial mass production in favor of artisanal integrity.

By the 1920s and 1930s, Art Deco introduced geometric symmetry, luxurious metals, and bold colour contrasts. Tables were set with angular glassware, chrome-plated cutlery, and mirrored trays that reflected light and created a sense of glamour. The 1925 International Exhibition of Modern Decorative and Industrial Arts in Paris cemented Art Deco as the dominant style, influencing everything from hotel dining rooms to middle-class homes.

Simultaneously, the rise of modernism, spearheaded by designers like Marcel Breuer and Ludwig Mies van der Rohe, advocated for "form follows function." Tables were stripped of unnecessary decoration; white linen cloths, unadorned porcelain plates, and clear glass tumblers became symbols of progressive thinking. This minimalist approach reflected the postWorld War I desire for efficiency and clarity, making table presentation a subtle yet powerful statement of modernity.

Mid-Century Modern: Functionality, Material Innovation, and Democratization

The period following World War II—particularly the 1950s and 1960s—ushered in the Mid-Century Modern movement, which profoundly reshaped table presentation. Characterized by clean lines, organic forms, and innovative materials, this era prioritized user experience and accessibility. Designers such as Ray and Charles Eames, Isamu Noguchi, and Tapio Wirkkala created tableware that was both sculptural and practical. The Eames' molded plastic dinnerware (introduced in the 1950s), for example, offered lightweight, durable, and stackable plates that revolutionized casual dining. New materials transformed the table. Melamine, developed in the 1930s and massproduced after the war, provided shatterresistant, colourful alternatives to traditional porcelain. Pyrex glassware, celebrated for its heat resistance, enabled elegant presentation of hot dishes without risk of breakage.

Stainless steel, perfected in the 1910s and widely adopted by the 1930s, replaced silver and pewter in cutlery, offering hygiene, durability, and a sleek modern aesthetic. Companies like Georg Jensen and Alessi elevated stainless steel to an art form, crafting ergonomic handles and polished finishes that balanced form and function.

The postwar economic boom democratized table presentation. Affordable, stylish tableware became accessible to the growing middle class. Casual dining emerged as a cultural norm, reducing the need for elaborate multicourse place settings. Instead, open, asymmetrical arrangements—featuring a few wellchosen elements such as a central vase, simple napkins, and coordinated plates—created harmony without formality. The use of placemats instead of tablecloths gained popularity, offering easy cleaning and a relaxed aesthetic. Colour palettes shifted toward earthy tones, pastels, and bold primaries, reflecting the optimism of the era.

Rise of Casual Dining, Buffets, and Thematic Presentation

The latter half of the twentieth century saw a significant shift toward casual dining, driven by changing social norms, the expansion of the restaurant industry, and the informality of domestic life. Formal, multicourse meals with intricate place settings gave way to simpler, communal arrangements. The buffet-style dinner, popularized in the 1960s and 1970s, demanded new approaches to table presentation. Long tables were arranged with food stations, each featuring coordinated serving dishes, decorative garnishes, and thematic elements. The emphasis was on visual appeal to stimulate appetite: colourful arrangements of fruits, vegetables, and pastries served as both decoration and sustenance.

Fig: Should you choose a Buffet or Sit-Down Dinner

In homes, the family-style meal became prevalent, with dishes placed at the centre of the table for communal sharing. This required tableware that was easy to handle and stack—such as stoneware bowls, wooden platters, and melamine trays. The counterculture movement of the 1970s further embraced rustic, handmade, and eclectic table settings, incorporating crafts, pottery, and natural elements like flowers, branches, and candles. This era celebrated individuality and rejected rigid formality, making table presentation more personal and expressive.

Simultaneously, themed dining gained popularity. Restaurants and event planners used table presentation to create immersive environments. Polynesian-themed establishments (e.g., Don the Beachcomber) featured tiki torches, bamboo placemats, and tropical centerpieces. Steakhouses employed rustic wood tables, leather napkins, and vintage-style plates to evoke classic American diner ambience. The Disney theme parks, opening in the 1950s and expanding through the century, elevated thematic dining with restaurants like The Crystal Palace, where characterthemed dishes and interactive table settings transported guests into a narrative experience.

Technological Advancements and New Materials

Innovation in materials science revolutionised table presentation throughout the century. Plastics, initially developed in the early 1900s, became ubiquitous after World War II. Melamine, introduced in the 1930s and popularised in the 1950s, provided lightweight, colourful, and virtually unbreakable tableware ideal for casual use. Acrylics and polycarbonates enabled transparent or translucent dishes that created modern, airy table settings. In glassware, tempered glass and borosilicate glass (as in Pyrex) allowed for thinner, more durable designs.

Finnish designers such as Timo Sarpaneva and Tapio Wirkkala experimented with glassblowing techniques to create sculptural, minimalist vessels that became centrepieces in themselves. The refrigeration revolution also impacted presentation. The ability to keep food fresh longer enabled elaborate cold presentations—charcuterie boards, salad bars, and chilled desserts—previously impractical. Food styling, emerging as a professional discipline in the midcentury, particularly with the rise of television cooking shows and food photography, introduced techniques such as arranging food in visually appealing ways, using edible garnishes, and employing props (specific plates or napkins) to enhance visual impact.

Globalization and CrossCultural Influences

Increased global travel, immigration, and cultural exchange in the twentieth century introduced diverse dining traditions to Western tables. The postcolonial era and the 1960s counterculture sparked interest in ethnic cuisines, which influenced table settings. Japanese sushi bars introduced sushi boats and minimalist bamboo mats; Indian restaurants incorporated brass thalis and colourful diyas (oil lamps); Mexican cuisine popularised handpainted ceramics and tortilla warmers. These elements served as cultural signifiers, enriching the dining experience. The fusion cuisine movement of the 1980s and 1990s further blurred cultural boundaries. A single table might feature sushi rolls on a Japanese lacquered tray, accompanied by Mediterranean mezze in Moroccan tagines, all arranged on a Scandinavian minimalist table.

This eclecticism reflected a globalised world where diners appreciated diverse aesthetics. International design movements also influenced Western tableware. Bauhaus principles inspired geometric glassware; Scandinavian design promoted natural wood and simple forms; Japanese wabisabi aesthetics celebrated imperfection and simplicity. These crosscultural influences created a rich tapestry of table presentation styles.

Thematic and Immersive Dining Experiences

Beginning in the latter half of the twentieth century, restaurants and event planners began using table presentation as a tool for creating thematic and immersive dining experiences.

Fig: Essential Tips for Memorable Themed Dinner

Themed restaurants—such as Polynesian lounges, steakhouses, and retro diners—designed entire environments, including table settings, to transport diners to another time or place. In fine dining, molecular gastronomy, pioneered by chefs like Ferran Adria of El Bulli in the 1990s, transformed table presentation into a scientific spectacle. Dishes were served in petri dishes, test tubes, or under cloches that released smoke or aroma when opened. The table became a laboratory, with precise plating and dramatic service enhancing the multisensory experience.

Even domestic gatherings embraced thematic presentation. Hosts organised Mediterranean nights, Mexican fiestas, or Victorian tea parties, coordinating tableware, linens, and decorations to create a cohesive atmosphere. This trend underscored that table presentation is not merely functional but also narrative—telling a story through arrangement.

Sustainability and EcoConscious Presentation

As environmental awareness grew, particularly from the 1970s onward, sustainability began influencing table presentation. The environmental movement prompted a shift toward ecofriendly materials. Reusable and biodegradable tableware—such as bamboo, bamboofiber composites, cornstarchbased plastics, and recycled glass—replaced disposable plastics. Restaurants adopted compostable napkins made from recycled paper or reusable cloth. The farmtotable movement, emerging in the 1980s and 1990s, emphasised locally sourced, seasonal ingredients, often presented in simple, rustic ways—wooden boards, slate plates, or ceramic bowls—to highlight natural beauty. Minimalist table settings, using fewer items and reusable materials, aligned with sustainable practices. The trend of zerowaste events encouraged permanent tableware, digital menus, and decor made from recycled or natural materials.Designers created elegant, sustainable tableware that married aesthetics with responsibility. Companies like Ecovaso produced stylish reusable items from ecofriendly materials, proving sustainability and beauty could coexist.

Digital Age: Technology at the Table

The late twentieth century introduced technology into table presentation. Digital design tools—such as Revit, SketchUp, and

3D modelling software—allowed precise planning of table layouts. Digital menus displayed on tablets replaced paper, reducing waste and enabling dynamic visuals. QR codes linked to wine pairings or dish stories enhanced guest engagement. Social media, particularly Instagram, turned table presentation into visual marketing. Restaurants invested in "Instagrammable" presentations—layered desserts, colourful arrangements—to encourage usergenerated promotion. In homes, smart home technology influenced settings: smart lighting adjusted ambiance; wireless speakers provided music without clutter. Smart tableware, such as temperaturecontrolled plates, kept food at ideal serving temperatures.

3

Materials, Fabrics, and Tools used in Napkin Folding

Classification of Napkin Fabrics and Textures

The classification of napkin fabrics and textures provides valuable insight into the intersection of functionality, aesthetics, and cultural practice within dining environments. By examining natural, synthetic, and blended fabrics alongside their textures, students and professionals gain a comprehensive understanding of how material choices influence both service quality and visual presentation. Whether applied in academic research or vocational training, this classification framework supports informed decision-making and reinforces the importance of textiles in shaping refined and effective table settings.

Importance of Fabric Choice in Napkin Design

The fabric used for napkins directly affects their performance, maintenance, and overall presentation. In academic and professional settings, fabric selection is closely linked to service standards and guest expectations. Fine dining establishments often prioritize elegance and texture, while institutional and casual dining environments focus on practicality and cost efficiency. Studying fabric types helps learners appreciate how material properties such as absorbency, strength, and drape determine the suitability of napkins for various purposes.

Natural Fiber Napkins and their Characteristics

Natural fiber napkins have long been valued for their comfort, absorbency, and traditional appeal. Fabrics such as cotton, linen, and silk are commonly discussed in academic curricula due to their historical significance and continued relevance.

Fig: Choosing the Right Napkin Material Makes a Significant Impact

Fig: Characteristics and Qualities of Napkins Made from Natural Fibers

Linen napkins, in particular, are associated with luxury dining and formal occasions because of their crisp texture and ability to hold intricate folds. Cotton napkins, by contrast, are versatile and widely used due to their softness, durability, and ease of care. These natural fabrics provide a tactile quality that enhances the dining experience while reflecting craftsmanship and tradition.

Synthetic Fiber Napkins in Modern Use

Synthetic fabrics have gained prominence with the growth of large-scale hospitality operations and the need for cost-effective solutions. Polyester and blended fabrics are frequently used in hotels, restaurants, and catering services because they are wrinkle-resistant, long-lasting, and easy to maintain.

Fig: Contemporary Utilization of Napkins Made from Synthetic Fibers

From an academic perspective, synthetic napkins are studied for their technological development and functional advantages rather than aesthetic refinement. Their textures are often smoother and more uniform, supporting consistency in presentation across high-volume service environments.

Linen napkins, in particular, are associated with luxury dining and formal occasions because of their crisp texture and ability to hold intricate folds. Cotton napkins, by contrast, are versatile and widely used due to their softness, durability, and ease of care. These natural fabrics provide a tactile quality that enhances the dining experience while reflecting craftsmanship and tradition.

Blended Fabrics and Functional Adaptability

Blended napkin fabrics combine natural and synthetic fibers to balance comfort, appearance, and durability. These blends are designed to retain the softness and absorbency of natural fibers while benefiting from the resilience of synthetics. In vocational training, blended fabrics are highlighted for their adaptability, as they suit both semi-formal and professional dining contexts. The study of blended textiles encourages learners to evaluate trade-offs between luxury and practicality in material selection.

Texture as a Sensory and Visual Element

Fig: Tactile, Olfactory, and Acoustic Elements in Architecture

Texture plays a crucial role in how napkins are perceived and used. Smooth textures convey formality and simplicity, while textured fabrics such as damask or waffle weaves add visual interest and tactile richness. Academic analysis often links texture to sensory perception, explaining how touch influences guest comfort and satisfaction. Textured napkins also interact differently with light and tableware, contributing to the overall aesthetic harmony of table settings.

Cultural and Contextual Influences on Fabric Selection

Cultural practices and dining traditions significantly influence napkin fabric and texture choices. Formal Western dining often favors linen or damask textures, while casual or contemporary settings may adopt cotton blends or disposable materials.

Fig: Textile Design Need to Know to Achieve the Best Results

In academic studies, these preferences are analyzed to illustrate how social norms, economic factors, and sustainability concerns shape material usage. This contextual understanding reinforces the idea that napkin classification is not purely technical but also cultural.

Table: Comparative of Napkin Fabrics and Textures

Fabric Type	Texture Characteristics	Common Usage
Linen	Crisp, smooth, structured	Fine dining and formal events
Cotton	Soft, smooth, flexible	Casual and everyday dining
Silk	Smooth, delicate, glossy	Decorative and ceremonial use
Polyester	Smooth, uniform, durable	Commercial hospitality
Blended Fabrics	Balanced softness and strength	Semi-formal and professional settings

Woven, Non-Woven, and Specialty Textures

Napkin textures can be further classified based on their construction methods. Woven napkins display distinct patterns and durability, making them suitable for repeated use and elaborate folding.

Non-woven napkins, commonly used in disposable settings, are designed for convenience and hygiene rather than longevity. Specialty textures, including embossed or jacquard designs, are often reserved for decorative or ceremonial purposes. Studying these variations helps learners understand how production techniques influence both texture and application.

Main Criteria for Classifying Napkin Fabrics and Textures

Mastering fabric may be vital, whether you're attempting to make new garments or find out how to clean your filthy ones. This is particularly the case if you wish to properly care for a fine piece of textile so that it lasts longer. Different fabric kinds have unique characteristics that might have a big impact on how you care for your clothes. And this is what derives fabric sourcing demand. For instance, the fiber composition of one fabric will have a completely different impact on how to clean the clothing than the fiber content of another fabric.

Let's examine the various types of fabrics to clear up some of the misunderstanding and get a better grasp of the textile world.

Various Types of Fabrics

First of all, a "fabric" is a substance created by weaving strands together. A fabric is often named after the fiber that was used to create it; some textiles may even have a combination of many distinct fibers. The fabric is then given a name based on the fiber(s) it was made of, its pattern and texture, and the manufacturing method utilized. Some textiles take the origin of the fibers into account. As a result, there are really two groups of classifications that first distinguish between the various forms of fabric: the types of fibers used (natural vs. synthetic) and the methods of manufacture (woven vs. knitted).

Fig: Natural and Synthetic

- Depending on the kind of fiber used, textiles can have a variety of initial differences in detail. The two categories are natural and manmade.
- Animals & plants both provide natural fibers. As an illustration, silkworms produce silk whereas plants produce cotton.
- On the contrary hand, man-made synthetic material only makes up synthetic fibers.

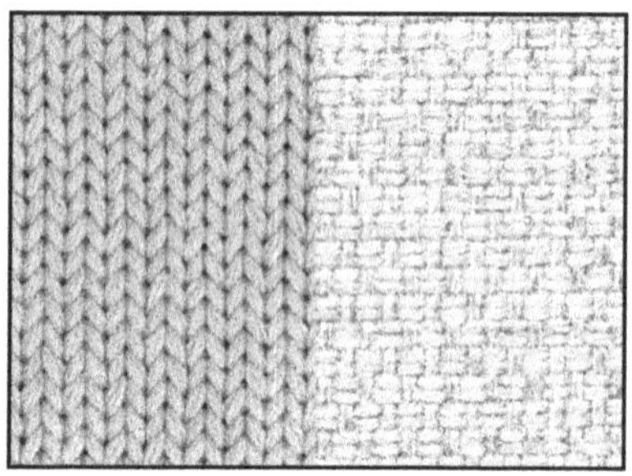

Fig: Woven and knitted

The production method employed is the second unique detail. Weaved and knitted are the two different kinds.

- Two components of yarn are intertwined horizontally & vertically on such a loom to create woven textiles. The material doesn't move because the yarn flows at a 45-degree angle, making it tauter and more durable than knit materials. A weft (when the yarn spans the cloth's breadth) and a warp make up the fabric when the yarn goes down the length of the loom.
- The three different weave kinds are twill, satin, or plain weave. Chiffon, crepe, denim, linen, satin, & silk are a few examples of common woven materials.

- Consider hand-knit scarves as an example of a knit material; the yarn is woven into a pattern of interlocking loops that enables the cloth to stretch considerably. Knit materials are renowned for their elasticity and form retention.
- Warp-knitted & weft-knitted fabrics are the two varieties of woven fabrics. Lace, lycra, and mesh are a few instances of common knit materials.
- Let's now examine the various fabric kinds.
- **Chiffon** is a sheer, thin, plain-woven material with a somewhat rough texture that is manufactured from twisting yarn. Typically, rayon, polyester, nylon, or silk are used to make the yarn.

Fig: Chiffon

- Chiffon may be readily coloured and is commonly seen in wraps, blouses and costumes, especially bridal gowns and prom outfits, due to its lighter, flowing texture.
- **Cotton** is one of most soft, natural looking fabrics that is regarded as the most widely used textile in the world. Ginning is the technique used to separate the fluffy fiber from cotton plant seeds. After that, the fiber is spun into fabric, which can then be knit or weaved.

Fig: Cotton

- The comfort, adaptability, and durability of this cloth are commended. It breathes nicely and is hypoallergenic, however it takes a while to dry. Shirts, dresses, and undergarments are all common examples of apparel that contains cotton. It may, nonetheless, crease and contract.

- **Chino, chintz, gingham, and muslin** are just a few of the various **fabric** kinds that cotton may produce.
- Delicate, tangled **plain-woven crepe** has a lumpy, uneven surface and doesn't crease. It is a flexible fabric that is frequently produced from cotton, silk, wool, or synthetic fibers. As a result, crepe is frequently referred to by the name of its material, such as crepe silk or crepe chiffon.

Fig: Crepe

- Due to its softness, convenience, and ease of usage, crepe is frequently one of most demanded wholesale fabrics used to make suits and dresses. For instance, the crepe fabric known as georgette is frequently used in fashionable clothing. Additionally, blouses, slacks, scarves, shirts, and skirts are made of crepe.
- **Denim** is a different sort of cloth. A cotton twill material called denim is created by weaving white cotton filling yarn and cotton wrapping yarn together. It is frequently praised for its colorful texture, durability, convenience, and endurance.

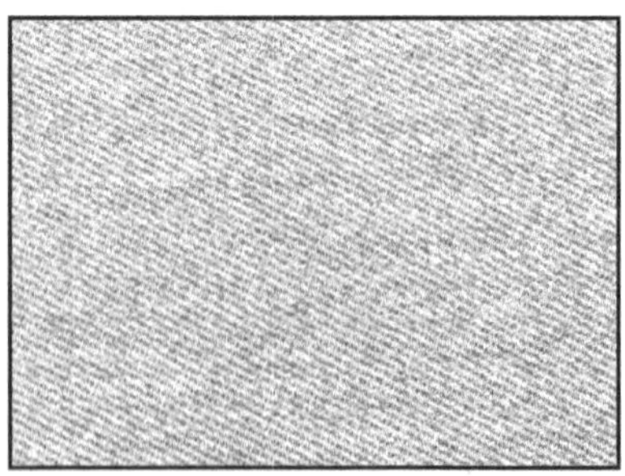

Fig: Denim

- Blue jeans apparel are often made of denim that has been indigo-dyed, although it may also be used to make jackets and outfits.
- In contrast to the other materials on this list, **synthetics** include spandex, nylon, and polyester. With delicate textiles, synthetic fibers don't shrink and often withstand water-based stains.

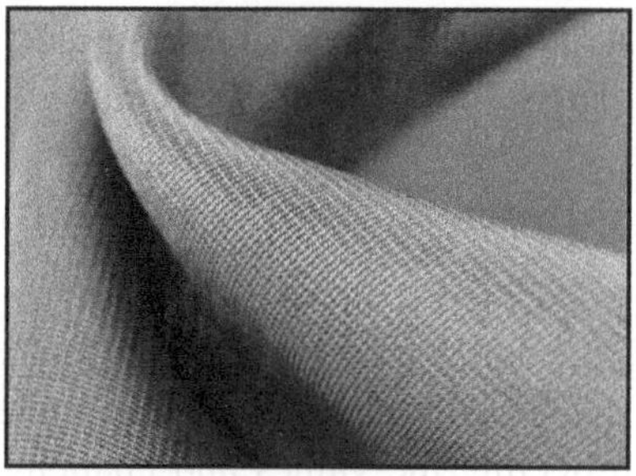

Fig: Denim

The classification of napkin fabrics and textures is guided by several important criteria that are emphasized in both academic and vocational studies.

- Fiber composition and origin.
- Degree of absorbency and softness.
- Durability and resistance to wear.
- Ease of washing and maintenance.
- Ability to hold folds and shapes.
- Visual appeal and surface finish.
- Suitability for formal or casual dining.
- Cost efficiency and availability.
- Environmental impact and sustainability.
- Compatibility with table décor themes.

Educational and Professional Significance

In educational settings, the study of napkin fabrics and textures strengthens learners' understanding of textile science, service standards, and aesthetic judgment. Vocational programs emphasize practical evaluation, allowing students to handle different fabrics and assess their performance in real service scenarios. This knowledge prepares learners to meet industry expectations, make cost-effective decisions, and maintain consistency in presentation. Academically, it also encourages analytical thinking by linking material properties to functional outcomes.

Linen, Cotton, and Synthetic Material Analysis

Linen, cotton, and synthetic materials are more than just fibers; they are threads in the tapestry of human history. Each has shaped cultures, economies, and environments in unique ways, reflecting the values and challenges of their respective eras. As we move forward, the choices we make about the materials we use will determine not only the future of the textile industry but also the health of our planet. In this complex weave, sustainability, innovation, and responsibility must guide the way, ensuring that the fabric of civilization remains strong, resilient, and beautiful.

As the world grapples with climate change and resource depletion, the textile industry is at a crossroads. Linen and organic cotton offer sustainable alternatives, but their production scales are limited. Synthetic materials, while problematic, continue to dominate due to their affordability and performance. The future of textiles lies in innovation: developing biodegradable synthetics, improving the efficiency of natural fiber production, and promoting circular fashion models that reduce waste.

Linen: The Noble Fiber of Antiquity

Linen, derived from the flax plant, is one of the oldest textiles known to humanity, with evidence of its use dating back over 10,000 years. Cultivated primarily for its fibers, flax requires minimal water and pesticides, making it an environmentally sustainable crop. The process of transforming flax into linen, however, is labor-intensive. It involves retting (soaking the stalks to separate fibers), scutching (removing woody parts), and hackling (combing the fibers). The result is a fiber that is strong, durable, and naturally lustrous. Linen's properties are unparalleled. It is highly breathable, making it ideal for warm climates, and it has natural antibacterial qualities. Its ability to absorb moisture without feeling damp has made it a staple in clothing, bedding, and tableware. Historically, linen was a symbol of purity and status, used in ancient Egypt for mummification and by European aristocracy for table settings and clothing. Its production declined with the rise of cotton and synthetic fibers, but it has seen a resurgence in recent years due to its sustainability and luxury appeal.

Cotton: The Fabric of the Industrial Revolution

Cotton, a soft, fluffy staple fiber, revolutionized the textile industry and became the backbone of the Industrial Revolution. Unlike linen, cotton is harvested from the boll of the cotton plant, a process that was mechanized with the invention of the cotton gin in the late 18th century. This innovation drastically increased production, making cotton accessible and affordable. However, the crop's rise was deeply intertwined with colonialism and slavery, as vast plantations in the Americas and India relied on exploited labor.

Cotton's versatility lies in its softness, breathability, and ease of dyeing. It is hypoallergenic, durable, and easy to care for, making it the most widely used natural fiber in the world. However, cotton cultivation is resource-intensive, requiring large amounts of water and pesticides. The environmental impact of conventional cotton has led to the rise of organic and sustainable cotton initiatives, which aim to reduce its ecological footprint.

Synthetic Materials: The Modern Revolution

The 20th century saw the advent of synthetic fibers, born from the petrochemical industry. Materials like nylon, polyester, acrylic, and spandex were developed to mimic or surpass the properties of natural fibers. Synthetic fibers are produced through chemical processes, such as polymerization, which transform raw materials like petroleum into long, continuous filaments. These fibers are engineered for specific qualities, such as elasticity, wrinkle resistance, and durability.

Polyester, the most common synthetic fiber, is lightweight, quick-drying, and resistant to wrinkles and shrinkage. It has become ubiquitous in fast fashion due to its low cost and versatility. However, synthetic materials come with significant environmental drawbacks. They are non-biodegradable, contributing to microplastic pollution in oceans and ecosystems. Additionally, their production relies on fossil fuels, releasing greenhouse gases and perpetuating dependence on non-renewable resources.

Environmental Impact: A Comparative Perspective

The environmental impact of these materials varies widely. Linen is the most sustainable of the three, requiring minimal water and chemicals, and flax plants improve soil health by returning nutrients to the earth. Cotton, while natural, has a substantial environmental footprint due to its water and pesticide requirements. Organic cotton mitigates some of these issues but remains less efficient than linen. Synthetic materials, despite their durability and low maintenance, are the least sustainable. Their production contributes to carbon emissions, and their disposal exacerbates pollution.

Cultural and Economic Significance

Each material reflects the cultural and economic values of its time. Linen, with its ancient origins, symbolizes tradition, luxury, and sustainability. It has been used in religious ceremonies, royal garments, and everyday items, embodying both sacredness and practicality. Cotton, the fabric of the Industrial Revolution, represents progress, accessibility, and globalization. It democratized clothing, making textiles available to the masses, but also perpetuated systems of exploitation. Synthetic materials, products of the modern era, embody innovation, convenience, and the tension between technological advancement and environmental responsibility.

Fabric Weight and its Impact on Fold Stability

Fabric weight plays a central role in determining how textiles fold, hold shape, and perform over time. Its interaction with gravity, stiffness, and drape defines the stability and precision of folds across a wide range of contexts. While heavier fabrics often provide greater structural support, their effectiveness depends on complementary properties and appropriate folding techniques. Lighter fabrics offer versatility and ease of handling but require additional reinforcement to maintain stable forms. By understanding the impact of fabric weight on fold stability, one gains deeper insight into the material logic behind textile practices and the enduring relationship between fabric properties and functional design.

Physical Properties Underlying Fabric Weight

Fig: Ultimate Guide to Understanding GSM in Fabrics

The weight of a fabric is closely connected to its physical structure, including fiber type, yarn thickness, and weave density.

Heavier fabrics typically contain thicker yarns or more tightly packed fibers, resulting in increased mass and reduced flexibility. This structural density affects how easily the fabric bends and how it responds to external forces such as pressure and gravity. Lighter fabrics, by contrast, are more responsive to movement and airflow, making them easier to manipulate but less capable of maintaining rigid folds. These underlying physical characteristics explain why fabric weight is a decisive factor in determining fold behavior and stability.

Relationship between Fabric Weight and Gravity

Gravity exerts a constant influence on folded fabric, pulling it downward and challenging the integrity of its structure. In heavier fabrics, the increased mass amplifies gravitational force, which can either enhance or undermine fold stability depending on the folding technique and fabric stiffness.

Dense fabrics with sufficient structural rigidity can resist collapse, allowing folds to remain upright and sharply defined. However, if the fabric is heavy yet overly flexible, gravity may cause folds to sag or spread, reducing their visual clarity. Lighter fabrics experience less gravitational pull, but their low mass often lacks the structural resistance needed to sustain complex or vertical folds.

Fabric Weight in Relation to Stiffness and Drape

While fabric weight is an important determinant of fold stability, it does not operate in isolation. Stiffness and drape are closely related properties that mediate the effect of weight on folding outcomes. Heavier fabrics often exhibit greater stiffness, which can support crisp, stable folds that maintain their form over time. This is particularly evident in starched linen or heavy cotton used in formal table settings. Conversely, some heavy fabrics with high drape, such as silk satin or viscose blends, flow easily and resist sharp folding despite their weight. In these cases, the fabric's tendency to conform to gravity outweighs the stabilizing influence of mass, resulting in softer and less durable folds.

Impact of Fabric Weight on Fold Precision

Fold precision refers to the clarity and sharpness of the folded edges and planes. Medium-weight fabrics often provide an optimal balance between flexibility and resistance, allowing for accurate folding with defined lines. Very light fabrics tend to blur fold lines due to their tendency to shift and wrinkle easily. Extremely heavy fabrics, on the other hand, may resist precise manipulation, making it difficult to achieve intricate or symmetrical folds. The relationship between fabric weight and precision highlights the importance of selecting appropriate textiles for specific folding designs, particularly in decorative or ceremonial contexts.

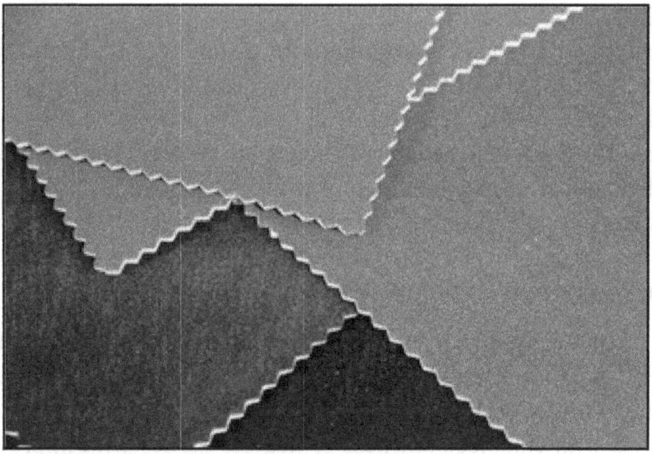

Fig: Fabric Weight in Order to Choose the Right Fabric

Durability of Folds Over Time

Fold stability is not only concerned with initial appearance but also with the ability of a fold to retain its shape over time. Fabric weight influences how folds respond to prolonged use, handling, and environmental factors such as humidity. Heavier fabrics generally offer greater resistance to deformation, enabling folds to remain intact during extended display. However, repeated folding and unfolding can create stress points in dense fabrics, leading to creasing or fiber fatigue.

Lighter fabrics may lose fold definition more quickly, especially in humid conditions, as moisture reduces fiber friction and increases pliability. These temporal considerations are essential in settings where folded textiles are expected to maintain their form for long periods.

Cultural and Practical Implications in Textile Use

The impact of fabric weight on fold stability has historically shaped cultural practices and aesthetic preferences. In formal dining traditions, heavier linens were favored for elaborate napkin folds because they conveyed elegance and permanence. In clothing and ceremonial textiles, weight influenced how garments were pleated and arranged, contributing to distinctive silhouettes and visual effects. Practical considerations, such as ease of laundering and storage, also affected fabric choices, balancing the desire for stable folds with the need for manageable textiles. These cultural and practical dimensions demonstrate how material properties influenced social customs and artistic expression.

Technological Developments and Fabric Engineering

Advancements in textile technology have expanded the range of fabric weights and their folding capabilities. Modern manufacturing techniques allow for the creation of lightweight fabrics with enhanced stiffness or heavier fabrics with improved flexibility. Finishing processes such as starching, resin treatment, and heat setting can modify fold stability without significantly altering fabric weight. These innovations have broadened the possibilities for folding design, enabling greater experimentation and consistency across different applications. The evolving relationship between fabric weight and engineered properties reflects the dynamic nature of textile science.

Color Theory and Pattern Considerations

Color theory and pattern considerations are inseparable pillars of visual design, each enhancing the other to create meaningful, aesthetically pleasing outcomes. By understanding the psychological impact of hues, the mechanics of pattern repetition, and the interplay between scale, contrast, and harmony, designers craft environments and products that resonate deeply with users.

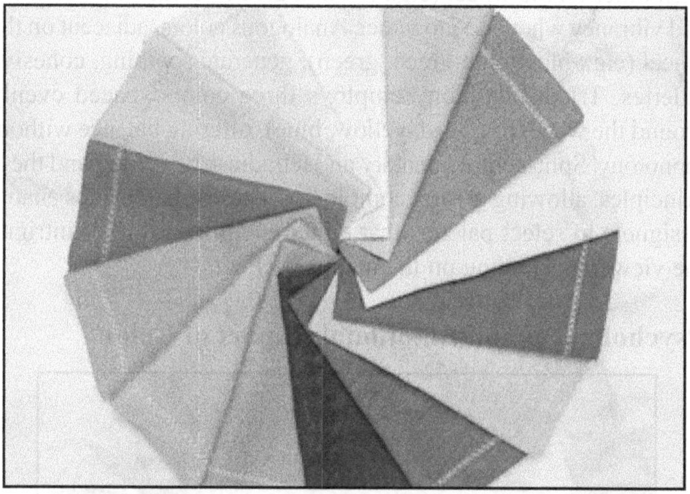

Whether in a tranquil bedroom adorned with soft blues and subtle stripes or a vibrant retail space bursting with complementary patterns, the deliberate application of these principles transforms ordinary spaces into extraordinary experiences. As design evolves, the timeless wisdom of color and pattern remains a vital compass, guiding creativity toward purposeful, beautiful results.

Fundamentals of Color Theory

At its core, color theory is a body of guidelines that explain how colors interact, combine, and affect human perception. The color wheel, developed by Isaac Newton and refined by subsequent theorists, serves as the primary tool for understanding these relationships. It organizes colors into three categories: primary colors—red, yellow, and blue—which cannot be created by mixing other colors; secondary colors—orange, green, and violet—formed by blending two primaries; and tertiary colors, which result from mixing a primary with a neighboring secondary (e://, red-orange).

Color harmony arises when colors are arranged in ways that produce visual appeal. Complementary colors, positioned opposite each other on the wheel (e.g., red and green), create high contrast and vibrancy when used together. Analogous colors, adjacent on the wheel (e.g., blue, blue-green, green), generate soothing, cohesive palettes. Triadic harmony employs three colors spaced evenly around the wheel (e.g., red, yellow, blue), offering balance without monotony. Split-complementary and tetradic schemes expand these principles, allowing greater complexity. These relationships enable designers to select palettes that either energize, calm, or intrigue the viewer, depending on the intended effect.

Psychological and Emotional Impact of Color

Fig: Folding Napkins "Leaves"

Beyond technical relationships, colors carry psychological and emotional weight that profoundly influences human behavior and mood. Red, for instance, is universally associated with energy, passion, and urgency—it stimulates appetite (hence its prevalence in restaurant branding) and signals danger (as in warning signs).

Blue evokes calmness, trust, and stability, making it a favored choice for corporate identities and healthcare environments. Yellow radiates optimism and warmth but, in excessive brightness, may cause visual fatigue. Green, linked to nature and growth, promotes tranquility and is often used in spaces intended for relaxation. Purple, historically rare and expensive, conveys luxury and creativity.

Cultural context further shapes color perception. In Western cultures, white symbolizes purity and is common in weddings, whereas in many East Asian traditions, it represents mourning. Black signifies elegance and sophistication in Western fashion but may carry negative connotations elsewhere. Designers must therefore consider both universal psychological responses and cultural nuances when selecting colors, ensuring their choices align with the target audience and context. This deliberate application of color psychology transforms spaces and products from merely functional to emotionally resonant.

Color in Different Contexts

The role of color varies dramatically across design disciplines, necessitating tailored approaches. In interior design, color establishes atmosphere, defines zones, and influences spatial perception. Dark tones can make a room feel intimate, while light colors expand perceived space. Accent walls introduce focal points without overwhelming. In fashion, color communicates identity, trend alignment, and seasonal themes—spring collections often feature pastels, while autumn leans toward earthy tones. Graphic design and branding rely on color to build recognition and convey corporate values; tech companies frequently adopt minimalist blue palettes to suggest innovation and reliability, while food brands use warm reds and yellows to stimulate appetite. Each context also demands consideration of color temperature—warm (reds, oranges, yellows) versus cool (blues, greens, purples)—and value (lightness or darkness). A well-balanced interior might pair a warm accent with cool neutrals to prevent visual fatigue. Similarly, a logo must maintain legibility across media, requiring sufficient contrast between text and background colors. These contextual adaptations demonstrate that color theory is not a rigid set of rules but a flexible toolkit applied with intention.

Pattern Considerations

Pattern—the repetition of a motif, shape, or design element—adds texture, rhythm, and visual complexity to surfaces. Unlike solid colors, patterns introduce movement and depth, guiding the eye across a space or object. Effective pattern design hinges on several key considerations: scale, repetition, contrast, and balance.

Scale refers to the size of the pattern relative to the surface it adorns. Large-scale patterns, such as oversized florals or geometric shapes, make bold statements and work well in spacious areas where they won't overwhelm. Small-scale patterns, like subtle dots or fine stripes, create delicate textures suitable for intimate settings. Repetition determines how frequently the motif recurs; tight repetition produces uniformity, while irregular spacing introduces organic, handcrafted appeal. Contrast between pattern and background ensures visibility—high-contrast pairings (e.g., black-and-white gingham) pop, while low-contrast combinations (e.g., pale gray on off-white) recede, offering subtlety.

Patterns fall into broad categories. Geometric patterns—stripes, chevrons, hexagons—convey order and modernity. Organic patterns, inspired by nature (leaves, flowers, waves), evoke softness and warmth. Abstract patterns use non-representational shapes to create visual interest without literal meaning. Textural patterns, such as embossed leather or woven textiles, add tactile dimension. The choice of pattern type depends on the desired aesthetic: a corporate office might employ clean geometric patterns for professionalism, while a boutique hotel lobby could feature botanical prints to evoke luxury and nature.

Integrating Color and Pattern

The synergy between color and pattern is where design truly comes alive. A poorly chosen color palette can render even the most beautiful pattern chaotic, while harmonious color application elevates simple motifs.

When integrating the two, designers must consider color dominance—whether the pattern contains multiple colors or relies on a single hue against a contrasting background. For instance, a multicolored paisley print requires a neutral base (white, beige, or gray) to prevent visual clutter. Conversely, a monochrome pattern—such as black-and-white houndstooth—gains impact when placed on a solid complementary color.

Color blocking within patterns—assigning specific hues to different elements of the motif—creates depth. A floral pattern might use analogous greens for leaves, a complementary red for blooms, and a neutral cream for stems, resulting in balanced vibrancy. Gradient patterns, which transition smoothly between colors, add sophistication and modernity.

Bunny Ear Napkin Fold

Nothing captures the playful spirit of mealtime quite like a bunny! This napkin fold is ideal for Easter, spring celebrations, or simply adding a bit of whimsy to your daily meals. The bunny ear napkin fold is easy enough for kids to master, yet charming enough to impress guests.

Fig: Create the Bunny Ear Napkin Fold

Materials Needed:

- Square striped napkins (preferably in pastel colors for a festive look)
- Napkin rings or thin ribbon (optional)

Step-by-Step Instructions:

- Lay the napkin flat on a table, making sure it's smooth and free of wrinkles.
- Fold the napkin in half to form a triangle.
- Start from the wide end of the triangle and roll the napkin tightly until you reach the point.
- Once rolled, fold it in half so that both ends point upwards, resembling bunny ears.
- Secure the base with a napkin ring or tie a small ribbon around it to hold the shape.
- **Kid-Friendly Tip:** Allow your children to pick their favorite colors for the napkins and ribbons. You can also let them add cute decorations like small stickers or paper cutouts to the napkin "ears" for extra flair.

Why Kids Will Love It:

- **Hands-on Creativity:** Kids will enjoy rolling and shaping the napkin to make it look like bunny ears.
- **Personalization:** They can decorate each napkin to give it a unique look, making them proud to show off their creations during dinner.
- **Buying Option:** For soft and durable cloth napkins, check out the collection at All Cotton and Linen. These high-quality napkins are perfect for all your folding needs and come in a variety of colors and patterns to match any theme.

Sailboat Napkin Fold

Does your child dream of sailing the high seas? Turn dinner into a nautical adventure with the sailboat napkin fold. This charming design is perfect for summer picnics, beach-themed dinners, or simply bringing a touch of the seaside to your kitchen table.

Fig: Create the Sailboat Napkin Fold

Materials Needed:

- Square border linen napkins (blue or white work great for a nautical theme)
- A napkin holder or small seashells for decoration (optional)
- Step-by-Step Instructions:
- Start with the napkin flat on a smooth surface.
- Fold it in half to make a triangle.
- Fold the left and right corners up towards the top point to create a diamond shape.
- Fold the bottom point of the diamond halfway up to form the base of the sailboat.
- Adjust the folds to make the napkin stand upright, resembling a sailboat.
- **Kid-Friendly Tip:** Let your kids get creative by adding little flags made from toothpicks and paper or decorating the sail with markers. This fold can also double as a place card holder for special occasions.

Why Kids Will Love It:

- **Interactive Play:** It's like crafting their own little sailboat! They can even use their boats to "sail" across the table.

- **Imagination:** This fold invites stories and games about sea adventures, encouraging kids to engage their imaginations during dinner.

Fancy Fan Napkin Fold

For a touch of elegance that even kids can enjoy, try the fancy fan napkin fold. This classic fold is perfect for special dinners, birthdays, or holiday feasts. The best part is, despite its sophisticated appearance, it's quite simple to create!

Fig: Create the Fancy Fan Napkin Fold

Materials Needed:

- Square plaid napkins (bright or patterned designs add an extra wow factor)
- A napkin holder or napkin rings to secure the fold
- Step-by-Step Instructions:
- Lay the napkin flat on a table.
- Start folding the napkin accordion-style, making small, even pleats from one end to the other.
- Once completely folded, hold it at the center and gently fan out the top edges.
- Place the bottom edge into a napkin holder or tuck it into a napkin ring to keep the fan shape.
- Kid-Friendly Tip: Teach your kids how to pleat the napkin evenly by showing them how to "pinch and fold." They'll love the hands-on nature of this task, and you'll get beautifully folded napkins in return!

Why Kids Will Love It:

- **Sense of Accomplishment:** This fold feels like a real accomplishment for kids. They'll be excited to show off their fancy creations to family members.
- **Adds Fun to Formal Meals:** The fan fold can make even a simple family dinner feel like a fancy occasion, making mealtime more enjoyable for everyone.

Fig: Napkin Folds to Make Mealtime Fun

Scale also interacts with color: large patterns benefit from restrained color palettes to avoid overwhelming the eye, while small patterns can tolerate bolder combinations due to their reduced visual weight. Additionally, pattern density—the amount of pattern per unit area—affects color perception. A dense pattern in high-saturation colors may feel aggressive, whereas a sparse pattern in muted tones feels airy.

Practical Applications

In interior design, pattern and color work together to define spaces. A living room might feature a bold, large-scale geometric wallpaper in teal and gold as an accent wall, balanced by solid neutral furniture and textured rugs. Curtains with a subtle stripe in complementary colors add rhythm without competing. In textile design, upholstery fabrics often combine patterns—such as a large floral print paired with a small-scale dot—using a coordinated color scheme to unify the look.

Graphic designers apply these principles to logos, packaging, and web interfaces. A logo using a repeating abstract pattern must maintain clarity at small sizes, necessitating high contrast and simplified forms. Packaging for a luxury product might employ metallic gold patterns on a deep burgundy background, leveraging color psychology to signal exclusivity.

In fashion, pattern and color dictate seasonal collections. A summer dress might feature a breezy, small-scale floral print in pastel pinks and blues on a white base, evoking freshness. A winter coat could use a large-scale tartan in rich burgundy, navy, and gold, conveying warmth and tradition.

Challenges and Solutions

Designing with color and pattern presents common pitfalls. Clashing colors—such as mismatched complementary hues lacking sufficient contrast—can create dissonance. Solution: use a color view filter or digital tools to test palettes. Pattern overload occurs when too many competing patterns crowd a space. The "60-30-10 rule" helps: 60% dominant color or pattern, 30% secondary, and 10% accent. Scale mismatches, such as pairing a tiny pattern with an oversized motif, disrupt harmony. Solution: maintain consistent scale relationships or use transitional elements (e.g., a medium-scale pattern) to bridge differences. Another challenge is lighting. Colors and patterns appear differently under natural versus artificial light. Designers should test samples under various lighting conditions. Additionally, printing or manufacturing limitations may alter intended colors; specifying Pantone codes or providing physical swatches ensures accuracy.

Sustainability and Color/Pattern

Modern design increasingly prioritizes sustainability, influencing color and pattern choices. Eco-friendly dyes—natural plant-based or low-impact synthetic pigments—reduce environmental harm. Designers may opt for earthy, muted palettes derived from sustainable sources, aligning with biophilic design principles. Pattern efficiency also contributes to sustainability. Digital printing allows on-demand pattern application, minimizing textile waste. Reusable patterns, such as modular designs that can be rearranged, extend product lifespans. Additionally, recycled materials—like repurposed fabrics featuring existing patterns—are celebrated for their unique character.

Fabric Care, Pressing, and Preparation Techniques

Fabric care, pressing, and preparation techniques are essential components of textile management that influence durability, hygiene, and presentation. By understanding fabric properties, adopting appropriate laundering and pressing methods, and applying systematic preparation routines, individuals and organizations can maintain high standards of textile quality. In academic and vocational studies, these techniques provide a practical framework for translating textile science into every day and professional applications, ensuring that fabrics continue to perform and appear at their best over time.

Significance of Fabric Care in Professional and Academic Settings

In professional settings such as hotels, restaurants, hospitals, and garment industries, fabric care is directly linked to service quality and brand image. Poorly maintained textiles can undermine professionalism and compromise user comfort. Academically, fabric care is studied as an applied discipline that combines chemistry, physics, and material science. Understanding how fibers react to washing, heat, and pressure allows students to make informed decisions about appropriate care methods. This knowledge is essential for developing systematic maintenance routines that preserve both aesthetic and functional attributes of fabrics.

Fabric Properties and Care Requirements

To choose the best fabric for napkins, it is important to have a good understanding of textile technology and functional finesse. There are several persuasive possibilities, such as cotton's commendable absorption capacity, linen's luxurious luster, and polyester blends' pragmatic prowess. To ensure an elegant and effective table setting, it is necessary to assess the texture, strength, and color of the fabric. Discerning diners demand napkins that are durable, delightful, and well-designed. Thus, choosing the right fabric is a crucial element of creating a perfect dining experience.

Fig: Best Fabric for Napkins—A Fabric Guide to Cloth Napkins

Effective fabric care begins with an understanding of fiber properties and construction. Natural fibers such as cotton, linen, wool, and silk respond differently to moisture, heat, and mechanical action. Synthetic fibers like polyester and nylon are more resilient but may be sensitive to high temperatures. Preparation techniques in academic training emphasize reading care labels, recognizing fiber blends, and anticipating fabric behavior during laundering and pressing. This foundational understanding ensures that care practices are tailored to material characteristics rather than applied uniformly.

When it comes to dining etiquette and table decorum, the choice of napkin fabric is a subtle yet significant element. It speaks to the attention to detail of the host and the quality of the dining experience. While it may go unnoticed by some, those who are well-versed in fine dining and hospitality recognize that the texture, appearance, and functionality of a napkin can either enhance or diminish the ambiance of the meal.

There are several options available for napkin fabrics. Each has its own unique properties. Whether you prefer the crispness of linen, the softness of cotton, or the practicality of synthetic blends, it is important to consider the characteristics of each material before making a choice. As we explore this topic further, it is also worth considering how factors such as environmental impact, care requirements, and cost efficiency might influence the decision-making process. Ultimately, choosing the best fabric for napkins can have a significant impact on the overall dining experience.

Key Takeaways

- **Absorbency** is an important factor to consider when choosing napkin fabric. It determines how efficiently the napkin can deal with spills and stains.
- **Durability** is crucial. Factors such as weave structure, fabric weight, and thread count contribute to a fabric's resilience to wear and tear.
- **Ease of cleaning** is another key consideration, with stain resistance, machine wash compatibility, and wrinkle propensity being important factors to evaluate.

- **Cotton**, **linen**, **bamboo**, and **polyester** are all viable options for napkin fabric. Each offers different benefits, such as softness, absorbency, durability, wrinkle resistance, and color retention.

Consider Several Factors When Choosing Napkin Fabric

Selecting the appropriate fabric for napkins necessitates evaluating several key factors.

Absorbency

When evaluating liquid absorption, it's crucial to assess the fabric's inherent absorbency. This ensures that it serves its purpose without leaving moisture on the diner's hands or table surface. In the realm of textile science, absorbency is a critical parameter. It defines the efficiency of a fabric in soaking up liquids and retaining them. The capillarity of fibers, porosity, and hydrophilicity are major factors influencing this attribute.

Durability

When evaluating napkin durability, one must consider the interplay of weave structure, fabric weight, and thread count. These elements collectively contribute to the textile's resilience to wear and tear as well as its ability to withstand repeated laundering. A critical analysis of these factors should guide the selection process. Durable fabrics offer longevity. They reduce the need for frequent replacements and, thus, prove cost-effective over time.

Ease of cleaning

Ease of maintenance is a critical consideration in the selection of napkin fabrics. This includes particular attention to the textile's stain resistance and its ability to withstand frequent laundering without a loss of texture or color. The following points highlight key aspects:

- **Stain Resistance**: Fabrics with tight weaves and specialized finishes enhance stain removal. These make them preferable for napkins that will be machine-washed regularly.

- **Pre-Wash Requirements**: Some fabrics may need a pre-wash to reduce future shrinkage and preserve fiber integrity during subsequent cleanings.
- **Machine Wash Compatibility**: The ability of the fabric to be safely machine-washed without deteriorating is vital for practicality and longevity.
- **Wrinkle Propensity**: Consider materials that do not wrinkle easily, thus reducing the need for laborious ironing after each wash.

A Guide to Choosing the Best Fabric for Napkins

Selecting the optimal textile for napkins needs an understanding of the intrinsic properties of various fibers.

Cotton

Cotton is a natural fiber characterized by its softness and high absorbency. It's suitable for both daily use and special occasions. Cotton's intrinsic properties make it an ideal selection for napkin fabric. It merges functionality with comfort. Quilting cotton, in particular, is prized for its durability and ease of maintenance. It's a top choice for crafting napkins.

- **Quilting Cotton**: Renowned for its sturdy yet soft texture, it is perfect for daily use.
- **Absorbent**: Exceptionally capable of handling spills and stains.
- **Soft**: Ensures a pleasant tactile experience for users.

- **Easy to Make**: Its forgiving weave accommodates even novice sewers.

Linen

Linen, characterized by its high-quality flax fibers, offers unparalleled elegance and durability. These make it a distinguished choice that caters to formal dining experiences. As a timeless classic in the world of table linens, fine linen showcases a lustrous finish that can be further enhanced by ironing. It adds a polished and refined touch to table settings. Its fiber structure is inherently soft and absorbent. It can efficiently blot up spills while providing guests with a gentle touch. But, a technical consideration is their propensity to shrink if not cared for properly. To ensure that these exquisite napkins remain timeless additions to any sophisticated table arrangement, it is crucial to adhere to specific laundering guidelines that maintain the integrity of linen fibers.

Bamboo

Bamboo fabric is becoming more and more popular as a material for napkins. This is due to its sustainable nature, softness, and absorbency. Bamboo is a standout choice. Because it's grown in an eco-friendly way, without the use of pesticides or synthetic fertilizers. Additionally, it has remarkable fiber properties.

- **Sustainability**: Bamboo's rapid regeneration and minimal environmental footprint resonate with eco-conscious consumers.

- **Luxurious Softness**: The tactile experience of bamboo fabric rivals that of high-end textiles, enhancing the dining experience.
- **Superior Absorbency**: Bamboo's microstructure allows for exceptional wicking and water retention. It can efficiently deal with spills.
- **Robust Durability**: Despite their supple feel, bamboo fibers exhibit high tensile strength. They offer longevity, even with regular laundering.

Polyester

Polyester is a synthetic fabric known for its exceptional durability and ease of maintenance. It stands as a cost-effective and versatile choice. It is suitable for a variety of dining settings.

- As a synthetic polymer derived from petrochemicals, polyester boasts high tensile strength and notable dimensional stability. It ensures that napkins keep their shape over time and through multiple wash cycles.
- Its inherent stain-resistant properties make it an ideal choice for environments where spills are commonplace. It can effectively repel liquids and reduce the risk of permanent damage.
- Additionally, polyester napkins are readily washable. They facilitate straightforward laundering without the need for specialized care.
- The availability of polyester in an array of colors and designs further enhances its appeal. It allows for the creation of both casual and elegantly appointed tablescapes.

Additional Tips

When setting up a dining table, it's important to choose napkins that complement the existing color scheme and patterns of the tablecloth or placemats. The fabric you choose can greatly impact the table's overall look and functionality. If you're interested in creating custom napkins, a tutorial on sewing cloth napkins can be incredibly helpful, especially for those who enjoy DIY projects. Consider these extra pointers, each designed to augment the dining experience:

- Opt for a double-sided fabric construction to ensure durability and a polished look from all angles.
- Use meticulous techniques to create mitered corners, which contribute to a refined and professional finish.
- Select a fabric with an appropriate edge treatment to withstand frequent laundering while maintaining its integrity.
- Explore the emotional resonance of color and pattern choices. These can subtly affect the dining atmosphere.

When creating custom napkins, it is important to take into account the weave, weight, and texture of the fabric. Understanding the properties of the fibers and how to properly care for the fabric is crucial to ensuring that the napkins not only look great but also last for a long time.

Laundering Practices and Fabric Preservation

Laundering is a critical component of fabric care that influences cleanliness, texture, and longevity. Appropriate washing methods prevent shrinkage, color fading, and fiber damage. In vocational studies, learners are trained to select suitable detergents, water temperatures, and washing cycles based on fabric type. Gentle agitation and proper rinsing are emphasized for delicate fabrics, while sturdier textiles require more robust cleaning.

Academic analysis of laundering explores the chemical interactions between detergents and fibers, reinforcing the scientific basis of practical routines.

Pressing Techniques and their Role in Fabric Appearance

Pressing, often confused with ironing, is a controlled technique that uses heat, moisture, and pressure to shape and smooth fabrics. In professional textile care, pressing is used to restore structure, create crisp edges, and enhance visual appeal. Academic instruction highlights the importance of temperature control, pressing direction, and the use of protective cloths to prevent scorching or shine. Effective pressing techniques improve fabric drape and finish, contributing significantly to overall presentation standards.

Preparation Techniques before and after Use

Fabric preparation extends beyond washing and pressing to include inspection, sorting, and storage. Before use, fabrics are examined for stains, damage, or manufacturing residues that may affect performance. After use, proper airing, folding, or hanging prevents odor retention and creasing. In hospitality and institutional environments, preparation techniques are standardized to ensure consistency and efficiency. Educational programs emphasize these routines as part of comprehensive textile management.

Equipment and Tools in Fabric Care and Pressing

The effectiveness of fabric care largely depends on the correct use of equipment and tools. Washing machines, dryers, steam irons, pressing tables, and fabric brushes each serve specific functions. Academic and vocational training includes instruction on equipment operation, safety precautions, and maintenance. Tool selection and usage helps learners achieve optimal results while minimizing fabric damage and energy consumption.

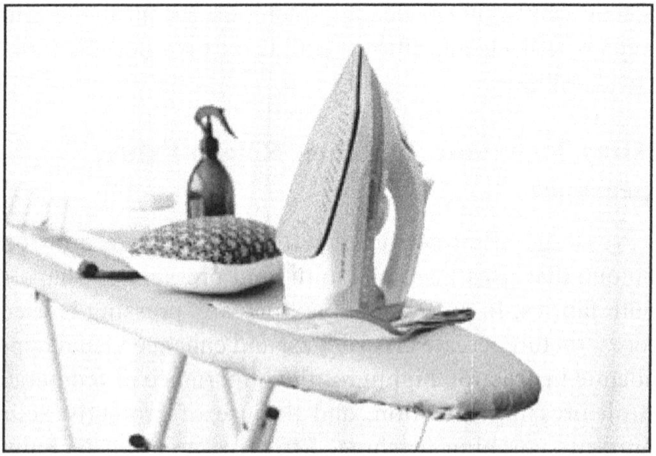

Fig: Tools for Professional Pressing

Comparative of Fabric Care and Pressing Requirements

The following table provides a comparative overview of common fabric types and their general care and pressing requirements.

Fabric Type	Washing Considerations	Pressing Requirements
Cotton	Warm water, moderate agitation	High heat with steam
Linen	Gentle wash, avoid overloading	High heat while damp
Wool	Cold water, minimal agitation	Low heat with steam
Silk	Hand wash or gentle cycle	Low heat with pressing cloth
Polyester	Warm water, quick drying	Medium heat, minimal steam

Principles of Effective Fabric Care and Preparation

Successful fabric care, pressing, and preparation are guided by several core principles that are emphasized in both academic and vocational learning environments:

- Understanding fiber composition and fabric structure.
- Following manufacturer care instructions accurately.
- Using appropriate water temperature and detergents.
- Applying correct heat and pressure during pressing.
- Preventing damage through protective tools and techniques.
- Maintaining cleanliness and hygiene standards.
- Ensuring consistency in professional textile handling.
- Incorporating inspection and quality checks.
- Adopting efficient storage and handling practices.
- Promoting sustainability through responsible care methods.

Educational Value and Skill Development

From an educational perspective, fabric care and pressing techniques cultivate discipline, attention to detail, and technical competence. Students develop practical skills that are directly applicable to industries such as hospitality, fashion, healthcare, and home management. Academic study encourages analytical thinking by linking observable outcomes to underlying material properties. This combination of theory and practice strengthens learners' ability to manage textiles effectively across varied contexts.

4

Principles of Design in Napkin Folding

Symmetry, Balance, and Proportion in Folding

Symmetry, balance, and proportion are the universal language of folding, transcending materials, cultures, and disciplines. They are the invisible threads that weave order and beauty into the act of transforming flat surfaces into three-dimensional forms. Whether in the delicate pleats of a couture gown, the precise creases of an origami masterpiece, or the functional folds of a solar panel, these principles remind us of the profound connection between mathematics, art, and human creativity. In mastering them, we not only create objects of utility and beauty but also participate in a timeless dialogue between order and chaos, form and function.

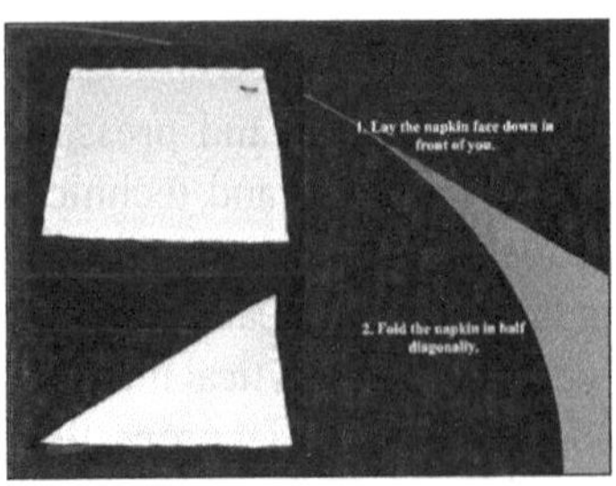

Fig: Table Napkin Folding Type 1

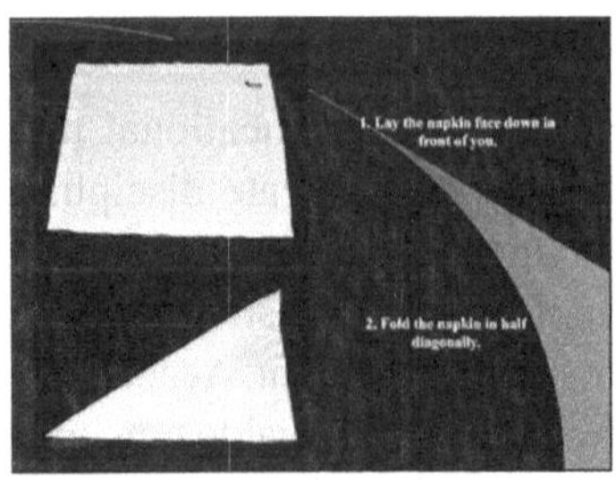

Fig: Table Napkin Folding Type 2

Symmetry: The Language of Order

Symmetry is the most immediate and recognizable principle in folding. It refers to the precise correspondence of parts on either side of a dividing line or plane. In folding, symmetry is often achieved through mirror-like reflections, rotational repetitions, or translational patterns. For example, a simple napkin fold in half creates bilateral symmetry, where one side is the exact mirror image of the other. This kind of symmetry evokes a sense of order and stability, appealing to the human preference for predictability and coherence.

In origami, symmetry is both a constraint and a creative challenge. Traditional models like the crane or the lotus rely on symmetrical folds to achieve their recognizable shapes. However, asymmetry can also be employed deliberately to introduce dynamism or realism. The choice between symmetry and asymmetry is not arbitrary; it is a deliberate decision that influences the emotional and visual impact of the folded object. Symmetry often conveys elegance and formality, while asymmetry can suggest movement or individuality.

Balance: The Dynamics of Equilibrium

Balance in folding goes beyond visual symmetry; it involves the distribution of weight, volume, and visual interest. There are two primary types of balance: static and dynamic. Static balance is achieved when the elements on either side of a fold are equal in size, shape, and visual weight, creating a sense of rest and stability. For instance, a table napkin folded into a perfect triangle exhibits static balance, with each side mirroring the other in mass and proportion.

Dynamic balance, on the other hand, introduces asymmetry while maintaining equilibrium. This is often seen in more complex folds where one side may be larger or more intricate, but the overall composition remains harmonious. In origami, dynamic balance is crucial for creating lifelike models, such as animals in motion. The folds must distribute the paper's mass in a way that suggests movement without collapsing the structure. This requires a deep understanding of the material's properties and the forces acting upon it.

Proportion: The Golden Ratio and Beyond

Proportion refers to the relationship between the sizes of different parts within a folded object. The most celebrated principle of proportion is the Golden Ratio, a mathematical concept where the ratio of two quantities is approximately 1.618:1. This ratio has been observed in nature, art, and architecture, and it can also be applied to folding. For example, a napkin folded into thirds rather than halves creates a more aesthetically pleasing rectangle, closer to the Golden Ratio.

In origami, proportion is critical for achieving realistic and visually appealing models. The size of a bird's wings relative to its body, or the length of a flower's stem compared to its petals, must adhere to natural proportions to be convincing. Proportion also dictates the scalability of designs. A fold pattern that works for a small sheet of paper may need adjustments in proportion to function on a larger scale.

Intersection of Symmetry, Balance, and Proportion

These three principles are not isolated; they intersect and influence one another in the act of folding. Symmetry often contributes to balance, as mirrored elements distribute visual weight evenly. Proportion ensures that the symmetrical parts relate harmoniously to one another and to the whole. For example, a symmetrically folded fan must have evenly spaced pleats (symmetry) that are wide enough to create a balanced spread (proportion) without overwhelming the table

Fig: Napkin Folds to Elevate the Look of Your Tablescape

setting (balance). In complex folding projects, such as modular origami or fabric pleating, the interplay of these principles becomes even more critical. Modular origami, where multiple folded units are assembled into a larger structure, relies on symmetry and proportion to ensure that the units fit together seamlessly. Fabric pleating, as seen in fashion or drapery, uses balance to distribute fullness evenly, creating a fluid and graceful silhouette.

Cultural and Historical Perspectives

The emphasis on symmetry, balance, and proportion in folding is not universal; it varies across cultures and historical periods. In Japanese origami, symmetry and proportion are highly valued, reflecting the culture's appreciation for precision and harmony. In contrast, some African and Indigenous American traditions embrace asymmetry and irregularity, viewing these qualities as more expressive and alive.

Historically, the Renaissance marked a turning point in the Western world's approach to these principles. Artists and architects began to codify the rules of proportion, as seen in Leonardo da Vinci's Vitruvian Man, which exemplifies the human body's ideal proportions. This period's influence can be seen in the meticulous folding of linen in European aristocratic traditions, where symmetry and proportion were used to convey status and refinement.

Practical Applications and Modern Innovations

Today, the principles of symmetry, balance, and proportion are applied in diverse fields beyond traditional folding arts. In engineering, foldable structures like solar panels and medical stents rely on precise proportions and balanced folds to function efficiently. In design, these principles inform the creation of foldable furniture, packaging, and even architecture, where buildings can expand or contract through innovative folding mechanisms. Modern technology, such as computer-aided design (CAD) and 3D printing, allows for the exploration of complex folding patterns that adhere to these principles with unprecedented precision. Algorithms can generate folds that achieve perfect symmetry, balance, and proportion, pushing the boundaries of what is possible with traditional methods.

Structural Geometry of Folded Forms

The structural geometry of folded forms provides the framework through which flat materials gain strength, dimension, and expressive potential. Through the strategic placement of creases, the careful management of angles, and the use of symmetry and repetition, folded structures achieve remarkable efficiency and resilience. Geometry guides how forces move through the form, how materials respond, and how visual coherence is maintained. By understanding these geometric foundations, one gains deeper insight into the enduring power of folding as a design strategy that bridges simplicity and complexity across cultures and disciplines.

Planar Surfaces and the Logic of Transformation

At the core of folded geometry lies the transformation of a planar surface into a volumetric structure. A flat sheet possesses minimal structural depth, but through folding, it gains rigidity and dimensional complexity. Geometrically, this transformation occurs through the introduction of creases that act as hinges, allowing planes to rotate relative to one another. Each fold alters the orientation of the surface, redistributing forces across the material. The placement and angle of folds determine whether the resulting form expands, contracts, or stabilizes, highlighting the importance of geometric planning in folded design.

Creases as Structural Elements

Creases are not merely lines of division but fundamental structural elements within folded forms. Geometrically, a crease defines an axis of rotation around which adjacent planes move. The angle and direction of this rotation influence the spatial configuration of the folded structure. In structural terms, creases concentrate stress and guide load paths, enabling the form to resist bending and compression. The repetition of creases in systematic patterns creates networks that enhance structural integrity, demonstrating how simple geometric interventions can produce complex and stable outcomes.

Angular Relationships and Fold Stability

The stability of folded forms is closely tied to the angular relationships between intersecting planes. Acute and obtuse fold angles produce different structural behaviors, affecting how forces are distributed throughout the form. Sharper angles often increase stiffness by limiting movement, while broader angles allow for flexibility and adaptability. Geometric consistency in these angles is essential for maintaining balance and symmetry. When angles are misaligned or irregular, stress concentrations can lead to deformation or collapse, underscoring the precision required in folded geometry.

Symmetry, Repetition, and Structural Efficiency

Symmetry and repetition play significant roles in the geometry of folded forms. Repeating geometric units, such as pleats or tessellated folds, distribute loads evenly and reduce material fatigue. Symmetrical arrangements enhance visual harmony while contributing to structural efficiency by balancing opposing forces. In many folded structures, repetition allows a single geometric principle to be scaled and extended, creating forms that are both expansive and cohesive. This interplay between symmetry and structure illustrates how geometric order supports functional performance.

Curvature and the Illusion of Flexibility

Although folds are often associated with straight lines, curved geometry also emerges within folded forms through the interaction of multiple planar segments. When a series of straight folds is arranged in a particular sequence, the resulting structure can approximate curvature, creating smooth, flowing surfaces. This geometric illusion of flexibility allows folded forms to adapt to complex shapes while retaining the advantages of planar construction. Curvature in folded geometry enhances load distribution and aesthetic appeal, bridging the gap between rigid structure and organic form.

Fig: Quick, Easy, Stylish Ways To Impress Dinner Guests

Material Constraints and Geometric Adaptation

The structural geometry of folded forms is inherently influenced by material constraints. Different materials respond uniquely to geometric manipulation, affecting fold sharpness, durability, and resilience. Geometry must adapt to these material properties, ensuring that fold patterns align with the material's capacity to bend without failure. In textiles, softer materials require more frequent folds to achieve structural stability, while stiffer materials can rely on fewer, more pronounced creases. This adaptive relationship highlights the necessity of integrating geometric design with material understanding.

Applications across Disciplines

Folded geometry extends beyond decorative or domestic contexts, influencing disciplines such as engineering, architecture, and industrial design. In architecture, folded surfaces create lightweight roofs and facades that combine strength with minimal material use. In engineering, folding principles inspire deployable structures and compact systems that expand efficiently. Across these applications, the structural geometry of folds enables innovation by transforming simple surfaces into multifunctional forms. The universality of folding geometry demonstrates its foundational role in shaping both practical and conceptual design solutions.

Temporal Behavior and Dynamic Geometry

Folded forms are not always static; their geometry often accommodates movement and transformation over time. Dynamic folding systems rely on geometric relationships that allow controlled motion without compromising structural integrity. Hinged folds, variable angles, and adaptive patterns enable forms to respond to changing conditions. This temporal dimension adds complexity to folded geometry, requiring designs that balance flexibility with stability. The capacity of folded structures to transition between states underscores the versatility of geometric principles in accommodating change.

Aesthetic Harmony with Table Settings

Aesthetic harmony in table settings is an art that marries design principles with cultural sensitivity and practical execution. By thoughtfully coordinating color, texture, scale, and arrangement, hosts transform a simple meal into a multisensory experience that delights the eye and nurtures connection. Whether through the precise symmetry of a formal banquet or the relaxed asymmetry of a family brunch, harmony ensures that every element feels intentional and interconnected. In doing so, it reflects care, creativity, and respect for the occasion and the guests. As design evolves, the timeless principles of balance, rhythm, emphasis, and unity remain the foundation for crafting tables that are not only functional but profoundly beautiful.

Principles of Design in Table Settings

At the heart of aesthetic harmony lie the fundamental principles of design: balance, rhythm, emphasis, and unity. Balance ensures that visual weight is distributed evenly across the table, preventing any single area from dominating. This can be achieved through symmetrical balance—mirroring items on either side of a central axis—or asymmetrical balance, where differing elements of varying visual weight offset one another. For example, a large floral centerpiece might be balanced by a cluster of candles and decorative plates on the opposite side.

Classic Pocket

- **Step One:** Start by laying a square dinner napkin flat.
- **Step Two:** Fold the top of the napkin a quarter of the way down.
- **Step Three**: Fold the bottom of the napkin up until it just covers the edge of the top of the napkin.
- **Step Four:** Flip the napkin over and fold into thirds.
- **Step Five:** Turn the napkin over again and place a name card, dinner menu, or flower into the pocket.

The Envelope

- **Step One:** Start by laying a dinner napkin flat.

- **Step Two:** Fold the napkin in half to form a triangle with the point facing away from you.

- **Step Three**: Fold the two bottom corners inward so that they meet in the center.
- **Step Four:** Fold the sides again toward the center, in thirds (approximately).
- **Step Five:** Fold the bottom up to create an envelope shape. Place a menu or note into the napkin, and fold down the top corner of the napkin.

The Drape

- **Step One:** Start by laying a square dinner napkin flat with the monogram on the underside of the bottom right corner.

- **Step Two:** Fold the napkin in half from left to right.
- **Step Three**: Fold the napkin in half again from bottom to top.
- **Step Four:** Turn the napkin so that the monogram is at the bottom and the napkin forms a diamond shape instead of a square.
- **Step Five:** Flip the napkin over and fold the top corner of the napkin one-third of the way down.

- **Step Six:** Fold the right and left corners of the napkin in at an angle so that the napkin forms a diamond shape.
- **Step Seven**: Turn the napkin over again and proudly display your monogram.

Double Diamond

- **Step One:** Start by laying a square dinner napkin flat.
- **Step Two:** Fold the napkin on a diagonal to create a triangle.
- **Step Three:** Fold the two bottom corners to meet the top, creating a square.
- **Step Four:** Carefully flip the napkin over and fold up the bottom corner to the desired length (about halfway or less).
- **Step Five:** Tuck under the left and right corners to create a more rectangular shape.

Angled Pocket

- **Step One:** Start by laying a square dinner napkin flat.
- **Step Two:** Fold the napkin in half from left to right.
- **Step Three:** Fold the napkin in half again from bottom to top.

- **Step Four:** Bring the top right corner of the first layer of the napkin down to the bottom left corner.
- **Step Five:** Flip the napkin over and fold into thirds.
- **Step Six:** Turn the napkin over again and place a name card, dinner menu, or flower into the pocket.

The Bow

- **Step One:** Start by laying a dinner napkin flat.
- **Step Two:** Fold the napkin on the diagonal to create a triangle.
- **Step Three:** Fold the top of the triangle one-third of the way down toward the center of the napkin, and fold the bottom up one-third toward the center.
- **Step Four:** Fold in half again so that the napkin is in the shape of a long strip. Press creases firmly.
- **Step Five:** With the napkin still horizontal, flip down both ends one-third of the way across the napkin so that they cross.
- **Step Six:** Using one hand to pinch the center of the napkin, slip on a napkin ring or ribbon to create the center of the bow.

Plate Wrap

- **Step One:** Start by laying a square dinner napkin flat.
- **Step Two:** Fold the napkin into thirds from top to bottom.

- **Step Three:** Flip the napkin over and place it on top of a salad plate, tucking the ends of the napkin underneath the plate.
- **Step Four:** Top with a name card, dinner menu, or flower.

Diamond Pocket

- **Step One:** Start by laying a square dinner napkin flat.
- **Step Two:** Fold the napkin in half so that the open end of the rectangle is facing toward you.
- **Step Three:** Fold the napkin in half again to create a square.
- **Step Four:** Rotate the napkin so the open corner faces away from you.
- **Step Five:** Take the open corner of the top layer and fold almost in half toward the bottom corner. Press.
- **Step Six:** Fold the next two layers as in the previous step, leaving a little space between corners to create a gradual effect. Make sure all layers look clean and even. Leave the last corner unfolded.
- **Step Seven:** Tuck the left and right corners underneath the napkin approximately one-third of the way on either side.

The Knot

- **Step One:** Start by laying a square dinner napkin flat.
- **Step Two:** Gather the top left corner and the bottom right corner of the napkin together in the center, making sure to tuck the points of the napkin into the middle of the bundle.
- **Step Three:** Tie the napkin in a knot and loosen it as needed for a more casual look.

The Leaf

- **Step One:** Start by laying a dinner napkin flat.
- **Step Two:** Fold the napkin on the diagonal to create a triangle.
- **Step Three:** Fold the top corner down toward the center of the napkin about two inches.
- **Step Four:** Accordion fold back and forth along the entire napkin in equal two-inch increments until you have a long strip. Press firmly.
- **Step Five:** Fold the napkin in half and match the two outer corners together while allowing the pleats to open, creating a leaf shape.

- **Step Six:** Add a napkin ring or ribbon at the bottom if desired.

Banded Angle

- **Step One:** Start by laying a square dinner napkin flat.
- **Step Two:** Fold the napkin in half from left to right.
- **Step Three:** Fold the napkin in half again from bottom to top.
- **Step Four:** Fold the top right corner of the first layer of the napkin down in one-inch increments until the fold reaches the top left and bottom right corners.
- **Step Five:** Repeat the previous step with the second layer of the napkin, folding down the corner in one-inch increments until it touches the first band.
- **Step Six:** Flip the napkin over and fold into thirds.
- **Step Seven:** Turn the napkin over again and place a name card, dinner menu, or flower into the pocket.

Pointed Pocket

- **Step One:** Start by laying a square dinner napkin flat.
- **Step Two:** Fold the napkin in half from left to right.
- **Step Three:** Fold the napkin in half again from bottom to top.
- **Step Four:** Turn the napkin so that the top right corner is at the top and the napkin forms a diamond shape.
- **Step Five:** Fold the top corner of the first layer of the napkin halfway down.

- **Step Six:** Flip the napkin over, and fold the left, right, and bottom corners of the napkin into the center until they touch.
- **Step Seven:** Turn the napkin over again and place a name card, dinner menu, or flower into the pocket.

Classic Fold

- **Step One:** Start by laying a square dinner napkin flat.
- **Step Two:** Fold the napkin in half from top to bottom.
- **Step Three:** Fold the napkin in half again, from left to right.
- **Step Four:** Fold the napkin in half a final time, from left to right.

Rhythm introduces repetition and variation, creating a sense of movement and continuity. Repeating a motif—such as a specific fold in napkins, a color in dishware, or a pattern in placemats—establishes rhythm without monotony. Emphasis directs attention to a focal point, such as a centerpiece or a statement piece of tableware, ensuring the eye has a natural resting place. Finally, unity binds all elements through consistent style, color, or theme, so that every item feels intentionally part of the whole. Together, these principles form the scaffolding for harmonious table design.

Color Coordination and Palette Selection

Color is arguably the most influential element in achieving aesthetic harmony. A well-chosen color palette unifies disparate items and sets the emotional tone of the meal. The palette should derive from a cohesive scheme—analogous (adjacent on the color wheel, e.g., blues and greens for a calming effect), complementary (opposite on the wheel, e.g., red and green for vibrancy), or monochromatic (variations in lightness and saturation of a single hue). Neutrals—such as ivory, beige, gray, or white—serve as versatile bases, allowing accent colors to shine without overwhelming.

Consistency is paramount: napkins, linens, dishware, floral arrangements, and decorative accessories should echo the same color family. For instance, a table set in soft blush pink might feature pale pink napkins, rose-gold cutlery, crystal glasses with pink stems, and a centerpiece of pink peonies. Metallic accents—such as gold, silver, or copper—can elevate a palette by adding reflective warmth. When introducing multiple colors, maintain a dominant, secondary, and accent ratio (e.g., 60% neutral, 30% secondary, 10% accent) to prevent visual clutter. Thoughtful color coordination ensures the table feels cohesive and visually restful.

Texture and Material Harmony

Texture adds depth, dimension, and tactile interest, preventing a table setting from appearing flat. Combining diverse materials—smooth porcelain, woven linens, glossy glass, matte ceramics, polished metal, and natural wood—creates contrast that enriches the aesthetic. However, harmony requires balance: an excess of competing textures can feel chaotic, while too little variation may seem sterile. A harmonious approach pairs contrasting textures in a deliberate way. For example, a crisp white linen tablecloth provides a smooth base, while textured woven placemats introduce tactile contrast. Polished silver cutlery gleams against matte stoneware plates, and a glass vase with smooth flowers adds reflective elegance. The interplay of textures should complement rather than conflict; a rustic wooden table might pair with burlap napkins and earthenware dishes, whereas a modern setting could feature sleek acrylic placemats, glossy porcelain, and metallic accents. Material harmony also considers weight—heavy crystal glassware balanced by lightweight linen—ensuring the table feels grounded yet airy.

Symmetry, Asymmetry, and Placement

The spatial arrangement of items on the table—whether symmetrical or asymmetrical—greatly influences visual harmony. Symmetrical placement involves mirroring elements on either side of a central axis, creating order, formality, and predictability. This approach is traditional in fine dining: identical place settings flank a centered floral arrangement, with cutlery aligned precisely. Symmetry conveys elegance and control, ideal for formal occasions.

Asymmetrical balance, conversely, relies on unequal elements that achieve equilibrium through thoughtful positioning. For example, a tall candle holder on one side might be balanced by a low, wide decorative bowl on the opposite side. Asymmetry offers flexibility and a contemporary, dynamic feel, suiting casual or eclectic gatherings. Regardless of the chosen style, consistency in the placement of key items is essential. The tops of plates, glasses, and cutlery should align horizontally, creating clean lines. The rule of thirds—dividing the table into three sections—helps position centerpieces and accessories without overcrowding. Proper spacing between place settings (typically 24–30 inches) ensures comfort and visual clarity.

Scale and Proportion

Scale—the relative size of objects—and proportion—their relationship to one another and the table—dictate whether a setting feels balanced or jarring. Oversized centerpieces may obstruct conversation or sightlines, while diminutive decorative pieces can be lost on a large surface. A general guideline is that a centerpiece should not exceed 12–14 inches in height to maintain clear views across the table. Proportional harmony extends to dishware stacking: a charger plate (largest) forms the base, followed by the dinner plate, salad plate (if used), and finally the bread plate (smallest). Cutlery should be arranged in the order of use, with each piece proportionate to its function—larger knives for main courses, smaller forks for appetizers. Glassware heights should also follow a logical sequence: water glasses (shortest) placed nearest the diner, wine glasses (taller) positioned behind. Maintaining proportional relationships ensures that no single element dominates, allowing the eye to move smoothly across the setting.

Lighting and Ambiance

Lighting is a silent yet powerful contributor to aesthetic harmony. It enhances colors, textures, and forms, shaping the mood of the dining experience. Natural light, when available, offers a soft, flattering glow that highlights details. However, artificial lighting—such as candles, pendant lights, or table lamps—allows for controlled ambiance.

Fig: Lighting and Ambiance —Napkin Folding Ideas

Candlelight casts a warm, intimate glow, ideal for romantic dinners, though it may create shadows that soften details. Overhead lighting should be diffused (e.g., using shades or dimmers) to avoid harsh glare on reflective surfaces. Accent lighting, such as LED strips beneath shelves or spotlights on centerpieces, adds drama without overwhelming. The color temperature of light also matters: warm white (2700K–3000K) enhances cozy, inviting settings, while cool white (3500K–4100K) suits modern, minimalist aesthetics. Integrating lighting with table design ensures that every element is visible and contributes to the intended atmosphere.

Role of Centerpieces

A centerpiece acts as the visual anchor of the table, drawing the eye and establishing the theme. Its design must harmonize with surrounding elements in scale, color, and style. Floral arrangements, candles, sculptural objects, or decorative bowls can serve this purpose, but their execution determines their effectiveness. For formal settings, a low, wide floral centerpiece in colors matching the table palette maintains elegance without obstructing views. In casual gatherings, a cluster of candles in varying heights or a simple fruit bowl adds charm.

The materials of the centerpiece should echo those elsewhere—a glass vase complements glassware, while a metallic base ties into silver cutlery. Additionally, the centerpiece should occupy roughly one-third of the table's length, leaving ample space for plates and conversation. A well-chosen centerpiece unifies the setting, reinforcing the theme and elevating the overall aesthetic.

Cultural and Contextual Sensitivity

Aesthetic harmony must respect cultural traditions and the context of the meal. Different cultures have distinct table-setting conventions that dictate placement, materials, and symbolism. In Japanese dining, minimalism prevails: low, natural wood tables, precise placement of chopsticks, and small, elegant dishes reflect wabi-sabi (appreciation of imperfection). Middle Eastern settings often feature colorful textiles, brassware, and communal platters, emphasizing hospitality. Western formal dining relies on symmetry, fine china, and multiple pieces of cutlery. Context—whether a wedding, holiday feast, business dinner, or outdoor picnic—also shapes decisions. A beachside gathering demands weather-resistant materials (bamboo, plastic, canvas) and casual arrangements, while a state dinner requires meticulous formality. Understanding these nuances ensures the table setting resonates with participants, honoring traditions and enhancing the experience through cultural familiarity.

Practical Strategies for Harmonious Table Settings

Creating aesthetic harmony is achievable with methodical planning. Begin by defining a unifying theme—such as a color scheme, seasonal motif, or stylistic direction (e.g., rustic, modern, vintage). Select linens, dishware, and decorations that align with this theme.

Use a place setting template:

- **Base layer**: Charger plate or placemat.
- **Dinnerware**: Dinner plate centered, salad plate (if used) atop it, bread plate to the upper left.
- **Cutlery**: Arranged in order of use, handles aligned, placed to the right (knives, spoons) and left (forks) of the plate.

- **Napkins**: Folded neatly and placed to the left of forks or in a napkin ring.
- **Glassware**: Positioned above the knives, tallest glass at the back.

Maintain consistent spacing between settings—approximately 24–30 inches—to ensure comfort. Test arrangements on the table before guests arrive, adjusting for balance and proportion. For large events, create a table plan to ensure uniformity across multiple tables. Finally, view the table from a distance to assess overall harmony, making refinements as needed.

Texture and Dimension in Fold Composition

Texture and dimension in fold composition are essential elements that elevate folded forms from simple arrangements to expressive, structured designs. Through the careful integration of surface quality and spatial depth, folded compositions achieve both aesthetic appeal and functional stability. In academic and vocational studies, this topic provides a rich framework for understanding material behavior, design principles, and practical execution. Mastery of texture and dimension ultimately enhances the ability to create folds that are visually engaging, structurally sound, and contextually appropriate.

Conceptual Foundations of Fold Composition

Fold composition refers to the deliberate arrangement of creases, layers, and planes to create a coherent form. From an academic perspective, it draws on principles of geometry, material science, and visual design. Texture contributes to surface quality and tactile experience, while dimension introduces depth, height, and volume. Together, these elements transform flat materials into structured compositions. Teaching fold composition often emphasizes how texture and dimension interact to guide the viewer's eye and support the stability of the folded object.

Role of Texture in Visual and Tactile Perception

Texture plays a critical role in how folded forms are perceived both visually and physically. Smooth textures create a sense of refinement and formality, while coarse or patterned textures add visual interest and complexity. In fold composition, texture affects how light is absorbed or reflected across surfaces, enhancing shadows and highlights that emphasize dimensional depth. Academic analysis often links texture to sensory perception, explaining how touch and sight work together to shape aesthetic judgment and user experience.

Dimension and Spatial Form

Dimension in fold composition refers to the transformation of a two-dimensional surface into a three-dimensional structure. Height, width, and depth are created through strategic creasing and layering. Dimensional folds provide structural integrity, allowing compositions to stand upright or maintain defined shapes. In vocational contexts, learners are trained to control dimension to achieve consistency and symmetry, while academic studies explore how dimensional variation communicates meaning, hierarchy, and emphasis within a design.

Interaction between Texture and Dimension

Texture and dimension are interdependent elements in fold composition. Textured surfaces can enhance the perception of depth by accentuating edges and planes, while smooth surfaces emphasize clean lines and geometric precision. The success of a folded composition often depends on balancing these elements so that neither overwhelms the other. Academic instruction highlights this interaction by encouraging experimentation with different materials and folding techniques to observe how texture influences dimensional stability and visual harmony.

Material Properties and their Influence on Folds

The material used in fold composition significantly affects both texture and dimension. Fabrics, paper, and other flexible materials respond differently to pressure, moisture, and manipulation. Thicker materials produce more pronounced dimensions but may limit intricate detailing, while lighter materials allow complex textures but require careful handling to maintain structure. Material behavior is a key component of academic and vocational training, linking theoretical knowledge with practical application.

Fig: Effects of Material Properties on Performance and Use

Aesthetic Balance and Structural Stability

Achieving aesthetic balance in fold composition involves aligning texture and dimension in a way that appears intentional and stable. Excessive texture can disrupt visual unity, while insufficient dimension may result in flat or unengaging forms. In professional settings such as hospitality and display design, balanced fold compositions contribute to polished presentations. Academic studies often analyze exemplary folds to demonstrate how proportion, symmetry, and repetition support both beauty and functionality.

Application in Design, Education, and Practice

Texture and dimension in fold composition have wide-ranging applications across disciplines. In hospitality education, they enhance table presentation and guest perception. In design studies, they support explorations of form and materiality. Vocational training emphasizes repeatable techniques that produce consistent results under time constraints. Across these contexts, fold composition serves as a practical exercise in translating abstract design principles into tangible outcomes.

Table: Texture and Dimension in Fold Composition

Element	Primary Function	Impact on Fold Composition
Texture	Surface quality and tactile appeal	Enhances visual interest and sensory engagement
Dimension	Creation of depth and volume	Defines structure and spatial presence
Smooth Surfaces	Visual clarity and refinement	Emphasizes symmetry and clean lines
Textured Surfaces	Pattern and complexity	Accentuates edges and layers
Layering	Structural support	Increases stability and depth

Considerations in Managing Texture and Dimension

- Effective fold composition relies on several important considerations that guide both learning and professional practice.
- Selection of appropriate material for desired texture.
- Control of crease placement to define dimension.
- Balance between visual complexity and clarity.
- Consistency of folds for symmetrical composition.
- Awareness of light and shadow on textured surfaces.
- Structural stability of dimensional forms.
- Adaptation of techniques to material thickness.

- Integration of aesthetic and functional objectives.
- Precision in execution to maintain form.
- Evaluation of overall harmony in the final composition.

Educational and Professional Significance

In educational environments, studying texture and dimension in fold composition fosters spatial awareness, creativity, and technical precision. Students learn to analyze how subtle changes in surface quality or depth alter perception and meaning. Professionally, these skills translate into improved presentation, design consistency, and attention to detail. The discipline required to manage texture and dimension effectively also cultivates patience and problem-solving abilities, which are valuable across creative and service-oriented fields.

5

Classification of Napkin Folding Techniques

Basic Folds and Foundational Techniques

The basic folds and foundational techniques of folding are simple in concept but profound in their applications. They are the alphabet of a universal language, allowing us to communicate ideas, solve problems, and create beauty across disciplines. Whether shaping paper, fabric, or even light, these fundamentals govern the transformation of flat planes into three-dimensional forms.

Mastery of these techniques is not just about technical skill; it is about understanding the principles of symmetry, balance, and proportion that underlie all folded forms. It is about seeing the potential in a single sheet of paper or a length of fabric and knowing how to coax it into being. In the end, the art of folding is a testament to human ingenuity, a reminder that even the simplest actions can lead to endless possibilities.

Valley Fold: The Gentle Crease of Creation

The valley fold is the most intuitive and fundamental fold, a soft bend that creates a V-shaped groove. It is the first lesson in any folding discipline, from paper crafts to fabric manipulation. To execute a valley fold, one simply folds the material toward the body, creating a crease that resembles a valley when viewed from above. This fold is the cornerstone of origami, where it is used to create the initial base for most models. In textiles, valley folds are the basis for pleats, darts, and tucks, shaping flat fabric into three-dimensional forms that drape and contour the body. The valley fold is not merely a physical action but a conceptual one. It introduces the idea of directionality in folding, as the material is guided along a specific path. This fold also teaches the importance of precision; a clean, sharp crease is essential for structural integrity and aesthetic appeal. In essence, the valley fold is the first step in transforming a two-dimensional plane into a world of possibilities.

Mountain Fold: The Inverse of Potential

The mountain fold is the valley fold's inverse, a bend that creates a ridge or peak when the material is folded away from the body. This fold is less intuitive than its counterpart, as it requires a mental flip to visualize the desired outcome. In origami, the mountain fold is often used to create layers, depth, and internal structures that support the model's shape. In textiles, it is employed in techniques like smocking and shirring, where the fabric is gathered and secured to create texture and volume.

The mountain fold introduces the concept of duality in folding. Every valley fold implies a mountain fold on the opposite side, and vice versa. This duality is fundamental to understanding how folds interact and how they can be combined to create complex patterns. The mountain fold also highlights the importance of flexibility in thinking; what appears as a peak from one side is a valley from another, a lesson in perspective that applies to both art and life.

Pleat: Repetition and Rhythm

The pleat is a series of alternating valley and mountain folds, creating a pattern of ridges and grooves. Pleats are among the most versatile and widely used folding techniques, appearing in everything from fashion to furniture design. In textiles, pleats can be knife-sharp or soft, wide or narrow, depending on the desired effect. They add structure, movement, and texture, allowing fabric to expand and contract gracefully. The pleat introduces the concept of repetition and rhythm in folding. Each fold must be consistent in width and depth to create a harmonious pattern. This requires patience and precision, as even a slight deviation can disrupt the overall effect. Pleats also teach the importance of planning, as the number and placement of folds determine the final shape and function of the object. In origami, pleats are used to create intricate patterns and organic forms, such as flowers or waves, demonstrating their potential for both utility and artistry.

Gather: The Art of Controlled Chaos

The gather is a technique where fabric or paper is bunched together along a single line, creating fullness and texture. Unlike the pleat, which is structured and uniform, the gather is loose and organic. In textiles, gathers are achieved by stitching along a line and then pulling the thread to create ruffles or poufs. In paper, gathers can be created by crumpling or folding in irregular patterns.

The gather introduces the concept of controlled chaos in folding. It allows for spontaneity and imperfection, qualities that can add warmth and character to a piece. However, even gathers require a degree of control; too much randomness can result in a messy or unbalanced outcome. The gather teaches the importance of knowing when to let go and when to rein in, a delicate balance that is central to the art of folding.

Twist: The Dance of Dimensions

The twist is a fold that turns one part of the material around another, creating a spiral or helical shape. This technique is less common than the valley or mountain fold but is incredibly powerful for adding movement and dynamism. In origami, twists are used to create intricate models like birds in flight or swirling patterns. In textiles, twists are employed in techniques like rope-making or creating bias-cut fabric spirals. The twist introduces the concept of three-dimensional manipulation in folding. It requires an understanding of how the material behaves when turned and how to control its tension and direction. The twist also highlights the importance of continuity; each turn must flow seamlessly into the next to create a cohesive form. This fold is a testament to the transformative power of folding, as it can turn a flat sheet into a dynamic, sculptural object.

Foundational Techniques: The Invisible Framework

Beyond the basic folds, there are foundational techniques that underpin all folding practices. These include creasing, scoring, and layering. Creasing involves creating a sharp, permanent fold, often using tools like bones or rulers. Scoring is similar but is typically done on thicker materials like cardboard or leather, where a groove is made to weaken the material along the fold line.

Layering involves stacking or folding material in multiple sheets to add thickness, strength, or complexity. These techniques are the invisible framework that supports the visible folds. They ensure that the material behaves predictably and that the folds remain crisp and defined. In origami, precise creasing is essential for achieving complex models, while in textiles, layering can add insulation or structure to a garment. These foundational techniques remind us that folding is as much about preparation and planning as it is about execution.

Intermediate Folding Structures

Intermediate folding structures play a vital role in the evolution of folded forms by bridging simplicity and complexity. They demonstrate how foundational folding principles can be expanded through layering, angular variation, and pattern repetition to create stable and expressive forms.

These structures balance aesthetic appeal with functional reliability, making them essential in both practical and artistic applications. By mastering intermediate folding structures, practitioners gain the skills and understanding necessary to explore more advanced folding systems, reinforcing the importance of this stage within the broader discipline of folded design.

Basic Folds to Structured Complexity

The transition from basic folds to intermediate structures involves the combination and sequencing of elementary folding actions. Simple folds such as single pleats or straight creases are repeated, layered, or intersected to create more dimensional forms. Geometrically, this stage introduces greater variation in angles, depths, and orientations, allowing flat materials to acquire volume and rhythm. These configurations demand a deeper understanding of spatial relationships, as each fold influences the behavior of adjacent planes. Intermediate folding structures thus demonstrate how complexity emerges not from entirely new techniques but from the strategic arrangement of fundamental folding units.

Layering and Overlapping in Fold Formation

Layering is a defining characteristic of intermediate folding structures. By stacking folded sections over one another, the material gains thickness and visual depth, enhancing both stability and aesthetic richness.

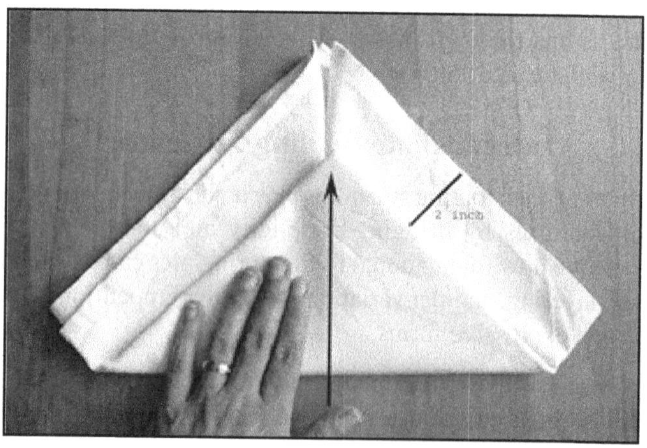

Fig: Fine Art of Napkin Folding

Overlapping folds distribute weight more evenly, reducing stress on individual creases and improving structural integrity. In textile applications, layered folds can support upright or semi-rigid forms, making them suitable for decorative or functional display. This layered geometry requires careful alignment to prevent distortion, highlighting the precision necessary at the intermediate level of folding practice.

Angular Variation and Spatial Control

Intermediate folding structures often incorporate varied angles to achieve more dynamic forms. Unlike basic folds, which typically rely on uniform angles, intermediate configurations use a combination of sharp and shallow folds to control spatial expansion and contraction. This variation allows folded forms to curve, spiral, or fan outward, creating visually engaging shapes. The management of angular relationships is critical, as inconsistent angles can disrupt balance and lead to instability. Mastery of angular control marks a significant step in the development of folding expertise.

Structural Balance and Load Distribution

As folding structures become more complex, considerations of balance and load distribution become increasingly important. Intermediate folding structures must support their own weight while maintaining intended form. This is achieved through symmetrical arrangements, evenly spaced folds, and strategic reinforcement at stress points. The geometry of the folds guides how forces move through the structure, preventing collapse or unwanted deformation. In practical applications, such as folded table linens or packaging, this balance ensures durability and consistent appearance during use.

Material Interaction and Responsive Folding

Material properties play a crucial role in shaping intermediate folding structures. At this level, the interaction between material flexibility, thickness, and surface texture becomes more pronounced. Materials that are too soft may require additional folds or layering to achieve stability, while stiffer materials demand careful handling to avoid cracking or uneven creases. Intermediate folding structures often exploit these material responses, using tension and compression to enhance form. This responsiveness underscores the importance of selecting appropriate materials and adjusting folding techniques accordingly.

Pattern Development and Repetition

Pattern formation is a key aspect of intermediate folding structures. Repetition of fold sequences creates rhythm and coherence, allowing complex forms to be built from recognizable units. These patterns may be linear, radial, or modular, depending on the desired outcome. Repetition not only contributes to visual harmony but also simplifies construction by establishing predictable relationships between folds. At the intermediate level, pattern development reflects an emerging design logic that bridges intuitive folding and systematic planning.

Functional and Decorative Applications

Intermediate folding structures are widely used in both functional and decorative contexts. In dining settings, they appear in structured napkin folds that balance elegance with practicality. In textiles and crafts, they are used to create pleated garments, decorative trims, and collapsible forms. These applications require folds that are visually refined yet manageable in terms of time and skill. The versatility of intermediate folding structures lies in their ability to enhance everyday objects without overwhelming complexity, making them accessible and adaptable across contexts.

Learning Progression and Skill Development

Engaging with intermediate folding structures represents a critical stage in skill development. Practitioners must move beyond mechanical repetition and develop spatial awareness, foresight, and precision. Mistakes at this level can propagate through the structure, making correction more challenging than in basic folds. As a result, intermediate folding encourages deliberate planning and thoughtful execution. This learning progression mirrors broader design processes, where increasing complexity demands greater conceptual clarity and technical control.

Advanced and Sculptural Fold Designs

Advanced sculptural fold designs crown napkin artistry, ephemeral monuments marrying craft with concept. From geometric rigor to organic reverie, they enchant transiently, dissolving into laps post-revelation. In design's continuum, they affirm folding's transcendent potential, tabling wonders wondrously.

Historical Progression to Complexity

Sculptural folds evolved from medieval mitres into 19th-century extravaganzas, Victorian banquets featuring feather boas of tiered ruffles that quivered realistically. Early 20th-century hotels codified peacocks with iridescent tail fans, pleats fanning radially from central spines. Post-war innovators like Lena Maslona cataloged over 200 forms, her "napkin bible" inspiring peacock trains that trailed elegantly. Japanese influences via noshi wrappings introduced angular cranes, wings extended impossibly from single sheets. This trajectory layered sophistication, each era adding pleat density and proportion finesse.

Geometric Abstractions and Modular Constructs

Pure geometry births crystalline forms, icosahedral polyhedrons emerging from mitred corners rolled into struts, self-assembling into geodesic domes over charcuterie. Hypercube projections fold flatly then inflate via hidden puffs; edges knife-sharp from interfaced linens. Modular systems interlock fan blades into kaleidoscopic mandalas, repeatable units tessellating banquets infinitely. Origami mathematicians inspire Miura folds expanding accordion-style into space-filling lattices, practical for voluminous centerpieces. These abstractions harmonize modern minimalism, stark whites or metallics gleaming architectonically.

Fig: Origami in Home Decor

Organic Mimicry in Floral and Faunal Sculptures

Organic sculptures mimic nature fluidly, rosebuds spiraling from accordion hearts, petals reflexing outward in graduated curls to unfurl progressively. Peacock displays radiate forty-pleat trains from hooded necks, quills tipped with beads for sway. Swan necks arch sinuously from hull bases, wings half-folded for lifelike poise. Orchid clusters multiply calyces via bifurcated tucks, stamen pistils protruding delicately. These biomorphic wonders demand supple fabrics, silks draping realistically where cottons stiffen structurally. Forms evolve mid-meal, swans' wings parting for sauce boats gracefully.

Structural Engineering for Endurance

Stability underpins sculpture, rolled dowels forming spines that cantilever wings aloft, counterweights in bases defying gravity. Interlocking tabs mimic dovetails, fan sections tabbing into hulls invisibly. Starch hierarchies grade rigidity, cores rigidified heavily, surfaces softly draped. Weighted hems anchor against drafts, filament threads invisibly lacing fragile extensions.

Engineering tests endurance, prototypes enduring buffet jostles before deployment. Precision irons pre-crease memory, folds springing true repeatedly.

Integrating Embellishments and Mixed Media

Embellishments amplify drama, LED filaments wiring swan eyes to flicker ethereally, or crystal droplets beading rose thorns sparkling. Fresh herbs tucked into lotus cores waft aromatically, edible sculptures blending scent with sight. Wire armatures subtly reinforce antler crowns for venison themes, disguised in fringe. Fabric paints airbrush gradients, ombré flames licking phoenix wings dramatically. Mixed media fusions embed origami paper birds within linen nests, hybrid fragility enthralling delicately.

Theatrical Deployments in Events

Theatrics shine in spectacles, processionals unveiling tiered fountains of cascading folds mid-banquet, synchronized reveals gasping audiences. Interactive installations invite guests pleating extensions, communal swans growing wingspans collectively. Projection-mapped sculptures shift colors dynamically, peacocks strutting virtual rainbows. Wedding arches of arched doves interlink overhead, folds fluttering in breezes romantically. Hospitality deploys kinetic mobiles, breeze-turned helices mesmerizing lobbies.

Technical Challenges and Artisanal Mastery

Challenges abound: oversized peacocks crumple under self-weight, solved by segmented construction. Asymmetric swans torque unstably, balanced via mirrored counter-folds. Humid climates wilt structures, mitigated by silicone sprays. Mastery accrues through repetition, fingers intuiting tuck sequences blindly. Apprenticeships in luxury liners hone speed, ninety-second roses routine. Digital mockups preview flaws, laser-cut guides accelerating novices.

Avant-Garde Interpretations Today

Contemporary artists reinterpret radically, Anish Kapoor-inspired voids sucking linen inward infinitely, negative space sculpting absence. Inflatable mylar folds balloon programmatically via apps, responsive installations pulsing breath-like. Biodegradable seafoam lattices dissolve post-feast, eco-sculptures vanishing gracefully. Gallery dinners feature wearable folds, diners donning plumed headdresses performatively.

Vertical versus Horizontal Fold Forms

Vertical and horizontal fold forms represent two fundamental approaches to fold orientation, each offering distinct advantages in terms of structure, aesthetics, and application. By understanding their conceptual foundations, structural characteristics, and contextual uses, learners and professionals can make informed decisions that enhance both visual appeal and functional effectiveness. In academic and vocational studies, the comparative analysis of these fold forms provides valuable insight into how simple directional choices shape perception, usability, and design coherence.

Conceptual of Vertical Fold Forms

Vertical fold forms are characterized by an upward or height-oriented structure that draws attention along a vertical axis. These folds emphasize elevation and prominence, often creating sculptural or architectural impressions. In napkin and textile applications, vertical folds are commonly used to add formality and visual drama to table settings. Academically, vertical fold forms are analyzed for their ability to create hierarchy and focal points, as they naturally guide the viewer's eye upward and establish dominance within a composition.

Structural Characteristics of Vertical Folds

The structural nature of vertical folds relies on layered

support and precise creasing to maintain upright posture. Stability is achieved through a strong base and balanced distribution of material. In vocational training, students learn that vertical folds demand accuracy and material awareness, as uneven creases or inappropriate fabric textures can cause collapse. The dimensional height of vertical folds enhances presence but also requires careful handling to ensure durability during use.

Fig: Fold a Linen Cocktail Napkin into a Minimalist Origami Crane

Conceptual of Horizontal Fold Forms

Horizontal fold forms extend laterally, emphasizing width, calmness, and visual continuity. These folds lie closer to the surface, creating a grounded and relaxed appearance. In dining and presentation contexts, horizontal folds are often associated with informality, accessibility, and ease. Academic discussions highlight how horizontal fold forms promote balance and visual stability by spreading visual weight evenly across the plane rather than concentrating it vertically.

Structural Characteristics of Horizontal Folds

Horizontal folds are generally more stable due to their low center of gravity and broader contact with the supporting surface. Their structure depends on smooth, evenly spaced creases that allow the material to rest naturally. In vocational practice, horizontal folds are favored for high-volume settings because they are quicker to execute and less prone to distortion. Despite their simplicity, effective horizontal folds still require attention to proportion and alignment to achieve a polished appearance.

Aesthetic and Functional Implications

The choice between vertical and horizontal fold forms has significant aesthetic and functional implications. Vertical folds convey elegance, formality, and celebration, making them suitable for ceremonial or fine-dining contexts. Horizontal folds, on the other hand, communicate comfort, simplicity, and approachability. From an academic perspective, these orientations are studied as visual language, where fold direction conveys mood, intent, and social cues within a setting.

Contextual Use in Education and Industry

In educational programs related to hospitality, design, and textile studies, vertical and horizontal fold forms are taught as complementary techniques rather than opposing choices. Students are encouraged to evaluate context, audience, and purpose before selecting a fold orientation. In industry practice, professionals apply vertical folds for special events and horizontal folds for everyday service, balancing efficiency with visual impact. This contextual understanding reinforces adaptability and design awareness.

Interaction with Material and Texture

Material type and texture significantly influence the effectiveness of both vertical and horizontal folds. Crisp fabrics with body, such as linen or starched cotton, support vertical folds by holding shape and height. Softer or more flexible materials are better suited to horizontal folds, as they drape naturally without requiring structural reinforcement. Academic instruction emphasizes testing materials to observe how texture and thickness affect fold orientation and stability.

Table: Comparative of Vertical and Horizontal Fold Forms

Aspect	Vertical Fold Forms	Horizontal Fold Forms
Visual Emphasis	Height and prominence	Width and balance
Structural Stability	Requires precise support	Naturally stable
Aesthetic Impression	Formal and dramatic	Casual and understated
Execution Complexity	More complex and detailed	Simpler and quicker
Common Applications	Ceremonial and fine dining	Every day and high-volume service

Considerations in Choosing Fold Orientation

Selecting between vertical and horizontal fold forms involves evaluating several practical and design-related factors:

- Purpose and formality of the occasion.
- Desired visual impact and focal emphasis.
- Material type and fabric stiffness.
- Available time and skill level.
- Stability requirements during use.
- Consistency across multiple settings.
- Compatibility with table size and layout.
- Integration with overall décor theme.
- Ease of handling and placement.
- Expectations of the target audience.

Educational and Professional Relevance

From an educational standpoint, the study of vertical versus horizontal fold forms develops spatial reasoning, design sensitivity, and technical proficiency. Students learn to analyze how orientation affects perception and functionality, linking theory with practice. Professionally, this knowledge supports efficient service, aesthetic consistency, and enhanced guest experience. Mastery of both orientations enables practitioners to adapt quickly to diverse contexts while maintaining high presentation standards.

Formal and Informal Folding Classifications

Formal and informal folding are not rigid categories but points on a continuum, each offering unique possibilities and challenges. Formal folding represents the pursuit of perfection, where every crease is a step toward an ideal form. Informal folding celebrates the beauty of impermanence, where the process is as important as the result. Together, they encompass the full spectrum of folding's potential, from the ceremonial to the mundane, from the mathematical to the intuitive.

These classifications, we gain insight into folding as a universal practice that bridges art, science, and culture. It is a testament to human creativity, a way of shaping the world around us with nothing more than our hands and a piece of material. Whether formal or informal, folding remains a powerful language, one that speaks of precision, expression, and the endless possibilities of transformation.

Formal Folding: Geometry, Precision, and Intentionality

Formal folding is the realm of exactitude, where every crease, angle, and proportion is deliberate and measured. It is governed by mathematical principles, often relying on symmetry, repetition, and the use of tools to achieve perfection. This classification is most evident in disciplines like origami, where diagrams and instructions dictate each fold with precision. The crane, for instance, requires a series of exact folds to transform a square sheet of paper into a recognizable bird. Similarly, in textiles, formal folding techniques such as pleating, darting, and box-folding are used in haute couture to create structured garments that maintain their shape.

Fig: Napkin Folding to Elevate your Napkin Ring Presentation

Formal folding is also deeply embedded in ceremonial and symbolic practices. The folding of flags in military ceremonies, for example, follows a strict protocol, each fold carrying a specific meaning. In Japanese tea ceremonies, the furoshiki (a traditional wrapping cloth) is folded with meticulous care to convey respect and elegance. These practices elevate folding beyond mere functionality, turning it into a language of honor, tradition, and artistry. The tools and materials used in formal folding further distinguish it. Rulers, bone folders, and templates ensure sharp creases and accurate measurements. Materials are often chosen for their ability to hold a fold, such as crisp paper or starched fabric. The result is a sense of permanence and authority, where the folded object stands as a testament to skill and discipline.

Informal Folding: Spontaneity, Utility, and Personal Expression

Informal folding is the antithesis of precision, embracing the organic and the impromptu. It is the folding of everyday life—the crumpling of a piece of paper, the casual draping of a cloth, or the haphazard tucking of a napkin. This classification is driven by practicality, where the goal is to achieve a functional result without adhering to strict rules. Informal folding is often intuitive, relying on tactile feedback and visual estimation rather than measurements. In textiles, informal folding is evident in techniques like gathering, shirring, and bundling.

These methods create texture and volume without the rigidity of formal pleats. A gathered skirt, for example, is achieved by loosely folding fabric along a thread, resulting in soft, undulating waves. Similarly, the art of wrapping gifts with recycled paper or fabric often involves informal folds, where the focus is on creativity and resourcefulness rather than perfection. Informal folding is also a medium for personal expression. Children instinctively fold paper into airplanes or boats, experimenting with shapes and forms without following a set pattern. Artists and designers draw on informal folding to create works that feel alive and dynamic, capturing the essence of movement and impermanence. This approach celebrates the human touch, with its inherent imperfections and uniqueness.

Cultural Contexts: Formality and Informality across Traditions

The distinction between formal and informal folding is deeply rooted in cultural traditions. In Japan, the line between the two is sharply drawn, with formal folding practices like origami and furoshiki existing alongside informal techniques used in daily life. The precision of formal folding reflects the Japanese values of discipline, respect, and attention to detail, while informal folding embodies the principles of wabi-sabi—finding beauty in imperfection and transience.

In Western cultures, the distinction is less rigid but still present. Formal folding is seen in the crisp napkin folds of fine dining or the structured draping of theatrical costumes, while informal folding dominates in domestic settings, such as folding laundry or packing a suitcase. However, there is a growing appreciation for the artistry of informal folding, particularly in contemporary design and fashion, where designers like Issey Miyake have elevated the informal into the realm of high art through techniques like pleating and draping.

Materiality and Technique: Defining the Divide

The materials used in folding play a significant role in determining whether the approach is formal or informal. Formal folding typically requires materials that can hold a sharp crease, such as heavyweight paper, cardboard, or starched fabric. These materials respond predictably to folding, allowing for precise control. Informal folding, on the other hand, often employs softer, more pliable materials like tissue paper, silk, or unstarched cotton, which lend themselves to fluid, organic shapes. Technique also defines the divide. Formal folding relies on predefined patterns and steps, often documented in diagrams or instructions. Informal folding is more improvisational, guided by the folder's intuition and the material's behavior. For example, folding a map to fit into a pocket is an informal act, driven by the need for practicality rather than aesthetics.

Intersection of Formal and Informal: Blurring the Lines

While formal and informal folding are distinct classifications, they are not mutually exclusive. Many practices exist at their intersection, blending precision with spontaneity. Modular origami, for instance, requires formal techniques to create individual units but allows for informal arrangement in the final composition. Similarly, in fashion, a garment may feature formal pleats alongside informal draping, combining structure with fluidity.

Contemporary artists and designers are increasingly blurring the lines between these classifications, using folding as a medium to explore themes of order and chaos, tradition and innovation. Works like those of artist Robert Lang, who combines mathematical precision with organic forms, or designer Ronan Bouroullec, who uses folding to create both functional and sculptural objects, exemplify this fusion.

Modular and Repetitive Folding Patterns

Modular and repetitive folding patterns exemplify the power of repetition as a design principle that unites structure, aesthetics, and functionality. By building complex forms from simple, repeatable units, these patterns achieve stability, scalability, and visual coherence. Their effectiveness across materials, cultures, and applications underscores their enduring relevance in folded design. Understanding modular and repetitive folding patterns provides valuable insight into how order and creativity coexist, demonstrating that repetition, when thoughtfully applied, can be a source of both strength and beauty in folded forms.

Concept of the Folding Module

A folding module is a self-contained unit created through a specific sequence of folds that can function independently while also connecting with identical units. Geometrically, the module establishes a set of proportions, angles, and crease relationships that determine how it interacts with neighboring units. The strength of modular folding lies in this consistency, as each unit contributes equally to the overall structure. By repeating a standardized module, designers can create extended surfaces or volumetric forms without introducing new folding rules, ensuring predictability and balance throughout the design.

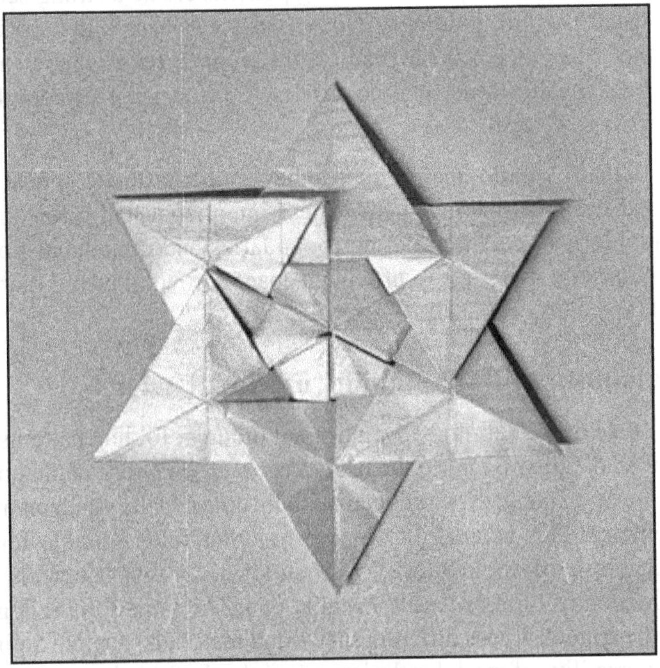

Fig: Folding Origami from Hexagonal Paper

Repetition as a Structural Strategy

Repetition in folding patterns is not merely decorative but serves as a critical structural strategy. When folding units are repeated, forces such as tension, compression, and gravity are distributed evenly across the structure. This uniform distribution enhances stability and reduces the likelihood of localized stress or deformation. In textiles and table linen, repetitive folds such as pleats or fans maintain shape through mutual reinforcement, where each fold supports the next. This structural logic explains why repetitive folding patterns are favored in applications that require durability and consistency.

Visual Rhythm and Spatial Order

Modular and repetitive folding patterns create a strong sense of visual rhythm, guiding the observer's eye across the surface of the folded form. The predictable recurrence of shapes and lines establishes order and coherence, even in large or complex structures.

This rhythmic quality contributes to the aesthetic appeal of folded designs, making them appear intentional and refined. In spatial terms, repetition defines boundaries and directional flow, allowing flat materials to articulate space through layered planes and repeating geometries.

Scalability and Adaptability of Modular Folds

One of the defining advantages of modular folding patterns is their scalability. Because the design is based on a repeatable unit, it can be expanded or reduced without altering its fundamental geometry. This adaptability makes modular folds suitable for a wide range of applications, from small decorative elements to large architectural or textile installations. The same folding logic can be applied across different scales and materials, demonstrating the versatility of modular design principles in folded structures.

Material Behavior in Repetitive Folding Systems

Material properties play a decisive role in the success of modular and repetitive folding patterns. Each repeated fold introduces stress into the material, making consistency in material thickness, flexibility, and surface finish essential.

Softer materials benefit from repetition, as the accumulation of folds increases rigidity and form retention. Stiffer materials require precise execution to prevent uneven stress distribution. The interaction between material behavior and repetitive geometry determines whether the folded pattern achieves the desired balance between flexibility and stability.

Cultural and Historical Contexts of Repetitive Folding

Repetitive folding patterns have deep cultural and historical roots, appearing in traditional textiles, ceremonial objects, and decorative arts across civilizations. In many cultures, repeated folds symbolized order, continuity, and harmony, reflecting broader philosophical and social values. These patterns were often transmitted through craft traditions, where repetition reinforced both skill mastery and cultural identity. The persistence of modular folding across time highlights its effectiveness as both a practical technique and a symbolic form of expression.

Complexity through Accumulation

Although modular folding patterns rely on simple units, complexity emerges through accumulation. As more modules are assembled, the overall structure gains depth, texture, and dimensional richness. This cumulative complexity allows folded forms to evolve organically, with each added unit contributing to the whole.

The designer's role shifts from creating individual folds to managing relationships between repeated elements, emphasizing systems thinking over isolated gestures. This approach mirrors broader design and architectural practices where complexity arises from systematic repetition.

Functional Applications and Practical Efficiency

In functional contexts, modular and repetitive folding patterns offer significant practical advantages. They simplify production by reducing variation, making folding processes more efficient and replicable. In domestic and commercial settings, such patterns are easier to standardize and maintain, ensuring consistent results. The predictability of modular folds also facilitates learning and skill transfer, allowing practitioners to refine technique through repetition while achieving reliable outcomes.

Fold Variations for Different Table Sizes

The relationship between napkin fold and table size is a nuanced dance of aesthetics, functionality, and intentionality. By tailoring folds to the dimensions and purpose of the table—whether through the compact triangle on a small bistro table, the elegant Water Lily on a medium gathering, or the towering Crown Fold adorning a grand banquet—hosts and designers craft settings that feel balanced, welcoming, and thoughtfully composed.

Each fold, chosen with consideration for scale, material, and context, transforms a simple square of fabric into a deliberate element of the dining experience. As the art of table presentation evolves, the principle remains: the fold must serve the table, not the other way around. In doing so, it elevates the mundane to the memorable, proving that even the smallest detail—like the way a napkin is folded—can shape the entire atmosphere of a meal.

Table Sizes and their Impact on Folding

Tables exist on a spectrum from intimate to expansive, each presenting unique challenges for napkin presentation. Small tables, typically defined as surfaces accommodating two to four diners—such as bistro tables, desk-style dining sets, or compact home tables—require folds that are unobtrusive, quick to execute, and spatially efficient. The limited surface area means that overly elaborate folds risk overwhelming the setting, competing with dishware, or obstructing the diner's reach:

- Medium tables, designed for groups of six to ten, offer greater flexibility. They allow for folds of moderate complexity that contribute to the table's visual rhythm without dominating it. Here, balance becomes key: the fold must complement—not overshadow—centerpieces, glassware, and plates.
- Large tables, intended for banquets, conferences, or formal dinners with twelve or more guests, demand folds that

are scalable, repeatable, and capable of creating unity across an expansive surface. A fold that appears delicate on a small table may vanish entirely on a large one, while an overly bulky design could disrupt sightlines or conversation.

Fig: Stand Design Table Napkin Folding for Dinnerware

Beyond physical dimensions, functional context influences fold choice. A busy café with small tables prioritizes speed and simplicity, whereas a formal event with large tables may allocate time for intricate, labor-intensive designs. Understanding these parameters is the foundation for selecting fold variations that respect the table's scale.

Fold Variations for Small Tables

For small tables, the guiding principle is minimalism. Folds must conserve space, remain easy to unfold, and avoid visual competition with the limited arrangement of dishes and decor:

- **The Triangle Fold is a Timeless Solution**: By folding the napkin diagonally into a neat triangle and tucking the pointed tip beneath, the result is a compact, stable form that sits gracefully beside the plate. Its low profile ensures it does not interfere with the placement of glasses or cutlery, which are often arranged in tight proximity on small surfaces. The triangle fold's simplicity also allows for rapid execution—a necessity in high-turnover settings like cafés or casual eateries.

- **The Pocket Fold offers Both Elegance and Utility**: The napkin is folded lengthwise into thirds, then the bottom edge is folded upward to create a small pouch. This pocket can discreetly hold utensils, toothpicks, or a menu card, consolidating items without spreading them across the table. Its flat, unobtrusive form makes it ideal for tables where every inch of space is precious.
- **For Modern, Minimalist Settings, the Half-Fold:** simply folding the napkin in half—provides a clean, accessible surface. Paired with a slender napkin ring, it adds a touch of refinement while maintaining simplicity. The roll fold, wherein the napkin is tightly rolled and secured with a ring, creates a cylindrical form that stands upright, conserving horizontal space and offering a contemporary aesthetic. In all cases, the chosen fold must align with the table's proportions, ensuring the napkin enhances rather than clutters the intimate setting.

Fold Variations for Medium Tables

Medium-sized tables grant designers greater creative latitude. Here, folds can introduce modest complexity while preserving harmony with the expanded arrangement of place settings, centerpieces, and decor:

- **The Bishop's Hat fold** excels in this context. Resembling a clerical hat, it stands upright with a peaked crown and broad base. Its verticality adds visual interest without consuming excessive horizontal space, making it suitable for tables where place settings are spaced farther apart. The fold's structured form also resists tipping, a practical advantage when diners may place items atop it.

- **The Water Lily fold,** which mimics the layered petals of a flower, brings organic elegance to medium tables. Beginning with a compact square base, the napkin is folded to create delicate, overlapping layers that splay outward. On a medium table, its moderate size ensures visibility and appreciation without dominating the setting. When paired with a low floral centerpiece, the Water Lily fold echoes the theme, reinforcing aesthetic cohesion.

For rustic or eclectic gatherings, the fan fold introduces movement and texture. The napkin is pleated into narrow strips, then fanned out to form a semi-circular shape. Placed beside the plate or used to cradle utensils, the fan fold adds dynamic visual rhythm. Its moderate scale prevents it from overwhelming the table while contributing to the overall decor. The crown fold, though more intricate, can be adapted for medium tables by simplifying its structure. Instead of a towering sculpture, a flattened crown—created by accordion-folding the napkin and tucking the ends to form a stable base—offers an elegant, manageable form. This fold works particularly well in semi-formal settings, where a touch of sophistication elevates the meal without excessive formality.

Fold Variations for Large Tables

Large tables present the challenge of unifying an expansive surface. Here, repetition, scalability, and boldness become essential. Folds must be visible across the table's length, contribute to a cohesive visual narrative, and avoid obstructing sightlines between guests:

- **The Crown Fold in its full,** sculptural form is ideal for large tables. Standing several inches tall, it serves as a visual anchor, drawing the eye down the table's length. When executed uniformly across all place settings, it creates a disciplined, impressive display. Its verticality also helps break up the horizontal expanse, making the table feel more intimate despite its size.

- **The Water Lily fold,** scaled slightly larger, maintains its delicate beauty while ensuring visibility. Its layered petals catch light and add texture, softening the table's scale. Paired with a low, wide floral centerpiece, the Water Lily fold enhances the theme without competing with other elements.

For efficiency in high-volume settings—such as conference banquets or wedding receptions—the roll fold with a napkin ring offers a practical solution. Tightly rolling the napkin and securing it creates a uniform cylinder that stands upright. On a large table, hundreds of identically rolled napkins establish order and consistency, reinforcing the event's formality. This fold is also forgiving; minor variations in execution are less noticeable at a distance.

Fig: Dressing up your Table

The tiered fold, which involves layering multiple napkins or creating a stepped effect within a single napkin, adds depth and dimension. This technique ensures the fold remains discernible even across a vast table. Luxury hotels and cruise lines frequently employ tiered folds to elevate the grandeur of formal dinners. Another option is the accordion fold, wherein the napkin is pleated lengthwise into a series of narrow strips. Arranged parallel to the table's long edge, these folds create rhythmic repetition that guides the eye along the table. This fold is particularly effective when the table features a long runner or linear centerpiece.

Adapting Folds for Irregular or Unconventional Tables

Not all tables conform to standard rectangles or circles. Irregular shapes—such as round tables, oval tables, modular arrangements, or tiered surfaces—require creative adaptations to maintain harmony:

- **For round tables,** folds that radiate outward complement the circular form. The fan fold or Water Lily fold, with their symmetrical, petal-like structures, mirror the table's curvature, enhancing cohesion. Placing these folds with their bases aligned to the table's edge ensures uniformity.
- **Oblong or oval tables** benefit from folds that follow the longitudinal axis. The accordion fold, created by pleating the napkin along its length, can be positioned parallel to the table's longer sides, reinforcing its geometry. Alternatively, the Bishop's Hat fold, oriented with its peak facing the table's center, creates radial symmetry.
- **Modular or sectional tables**—common in modern event design—allow for mixed fold styles. One segment might feature sculptural folds, while another employs simple rolls, provided a unifying element (e.g., consistent color, fabric, or napkin ring style) ties them together. This approach adds visual interest without chaos.
- **Tiered tables** (with multiple levels) demand height-sensitive adaptations. On lower tiers, slightly larger folds ensure visibility; on upper tiers, compact forms prevent obstruction. The triangle fold or half-fold work well on elevated sections, while lower tiers might showcase the Crown Fold.

Practical Considerations: Material and Setting Context

The choice of fold must also account for napkin material and setting context:

- **Linen Napkins,** prized for their crispness, hold sharp creases and are ideal for intricate sculptural folds like the Crown or Bishop's Hat. Their structured nature ensures designs maintain their form throughout the meal.
- **Cotton Napkins** offer softer, more casual aesthetics, suited to rustic folds such as the fan or pocket. Their flexibility allows for gentle curves but may require additional securing (e.g., napkin rings) for complex designs.
- **Paper Napkins,** especially in commercial settings, necessitate simpler folds due to lower structural integrity. The triangle fold, half-fold, or roll fold are reliable choices, as they minimize tearing and remain functional.
- **Setting Context further Refines Fold Selection:** In casual environments (e.g., family-style meals), practicality dominates: folds must be quick to execute and easy to unfold. In fine dining, time-intensive sculptural folds justify the effort, signaling luxury and attention to detail.
- **Outdoor Events introduce Weather Considerations:** Heavy linen or cotton napkins resist wind better than lightweight paper. Folds like the roll or triangle, secured with rings, minimize flutter.
- **Lighting also Plays a Role:** under candlelight, reflective folds (e.g., the Bishop's Hat) catch ambient glow, enhancing ambiance. In brightly lit spaces, bolder folds with high contrast ensure visibility.

6

Cultural and Regional Variations in Napkin Folding

European Traditions in Formal Table Presentation

European traditions in formal table presentation represent a timeless synthesis of history, etiquette, and aesthetic sensibility. From the geometric precision of French state dinners to the warm abundance of Italian feasts, these practices transform a meal into a ceremonial act that honors guests, reinforces social bonds, and celebrates cultural heritage.

Though modern influences have introduced adaptations—sustainability, technology, minimalist design—the foundational principles of balance, order, and intentionality remain unwavering. As societies evolve, so too does formal table presentation, ensuring that this cherished tradition continues to grace tables across Europe and beyond, a silent yet eloquent testament to the art of hospitality.

Historical Foundations of European Formal Dining

The origins of formal table presentation in Europe trace back to the Middle Ages, when feasting served as a public display of wealth, power, and hospitality among the nobility. Early medieval banquets featured communal trenchers (thick slices of bread used as plates) and shared bowls, but by the Renaissance, the introduction of fine ceramics, silverware, and linen tablecloths marked a dramatic evolution. Italian courts, particularly under the influence of the Medici family, pioneered elaborate settings that emphasized symmetry and abundance. The French court of Louis XIV further codified these practices, establishing service a la française—a style where all dishes for a course were presented simultaneously—demanding meticulously arranged tables. Etiquette manuals, such as Baldassare Castiglione's The Book of the Courtier (1528) and Antoine de Courtin's The Rules of Civility (1671), formalized behavioral norms and table settings, prescribing everything from the placement of cutlery to the manner of napkin use.

These texts reinforced the idea that dining etiquette was inseparable from social grace. By the 18th and 19th centuries, industrialization made fine tableware more accessible, yet the principles of formal presentation remained the preserve of the elite, solidifying traditions that continue to influence contemporary practice.

Role of Etiquette and Social Hierarchy

In European formal dining, etiquette is inextricably linked to social hierarchy. Historically, the arrangement of a table communicated the host's status and the relative importance of guests. The high table, positioned at the head of the room, was reserved for the most honored individuals—monarchs, dignitaries, or the host—while lower-ranking guests occupied peripheral seats. Even the placement of utensils and glassware followed strict protocols: the most prestigious guests received the most elaborate settings, often featuring multiple pieces of cutlery and crystal stemware. Etiquette manuals emphasized that every gesture—from the unfolding of a napkin to the sequence of courses—reinforced social order. For instance, a guest failing to align their cutlery precisely or misplacing a glass could be interpreted as a breach of respect. Though modern society has softened rigid class distinctions, the underlying principles persist. Contemporary formal settings still prioritize symmetry, order, and meticulous arrangement, ensuring that every element signals respect for both the host and fellow diners.

Elements of Formal Table Setting

A traditional European formal table setting is a study in

balance, precision, and harmony. Each component is placed with deliberate intention, creating a cohesive visual narrative.

Fig: Different Napkin Folds that'll Instantly Dress Up your Table Settings

- **Linens and Tablecloths**: The foundation is a crisp, white linen or cotton tablecloth, symbolizing purity and sophistication. In especially formal contexts, damask—a woven fabric with intricate patterns—may be used. Placemats, when employed, are rectangular and aligned perfectly parallel to the table's edge. The choice of linen reflects both tradition and luxury, as high-quality fabrics have long been markers of refinement.
- **Dishware and Glassware**: Formal settings feature layered dishware. At the base lies the charger plate, a decorative, oversized plate that anchors the place setting. Upon it rests the dinner plate, with a smaller salad plate placed atop for the first course. Above the plates, glassware is arranged in descending order of height. The water glass sits closest to the diner, followed by wine glasses—typically white wine, red wine, and dessert wine—positioned diagonally behind. Crystal glassware, prized for its brilliance and

clarity, is preferred, as its refractive quality enhances the table's elegance.

- **Cutlery Arrangement**: Cutlery follows a strict sequence of use, with utensils for the first course placed outermost and progressing inward. Forks are aligned to the left of the plate, knives and spoons to the, with knife blades facing the plate. The butter knife rests diagonally across the bread plate, which is positioned to the upper left of the setting. Dessert spoons and forks may be placed above the plate or introduced with the relevant course. Handles must be perfectly aligned, and spacing between items is uniform—typically 1–2 centimeters—to maintain visual harmony.
- **Napkins and Folding**: The napkin, traditionally folded into a simple rectangle or triangle, is placed to the left of the forks or secured in a napkin ring. In highly formal settings, it may be folded into a modest sculptural form—such as the Bishop's Hat—though excessive ornamentation is generally avoided. The napkin serves a dual purpose: functional (cleaning hands and mouth) and symbolic (signifying the meal's commencement when unfolded).
- **Centerpieces and Decor**: A centerpiece provides a focal point without obstructing conversation. European tradition favors low, wide floral arrangements in crystal vases, allowing guests to see one another across the table. Candles in polished silver or crystal holders add ambient light and a touch of romance; they are never placed directly in front of a guest. Seasonal motifs—such as holly for winter or spring blossoms—may adorn the table, reinforcing the occasion's theme.

Regional Variations within Europe

While formal table presentation shares universal principles, regional nuances reflect cultural identities.

- **In France, the epicenter of haute cuisine,** settings emphasize minimalism and mathematical precision. The service a la russe—where courses are served sequentially—demands multiple sets of glassware and cutlery, each placed with exacting symmetry. French tables often feature porcelain from renowned manufacturers like Limoges or Sèvres, paired with crystal from Baccarat or Lalique.
- **British formal dining**, influenced by royal tradition, incorporates subtle distinctions. A finger bowl—a small glass of water for rinsing fingertips—may appear with the final course. Afternoon tea settings include a three-tiered stand with scones, sandwiches, and pastries, accompanied by silver tea sets and bone china.
- **German settings favor substantial,** handcrafted ceramics and sturdy glassware, reflecting a preference for durability. Beer glasses may accompany wine glasses at traditional occasions, and bread plates are often larger to accommodate hearty Brot.
- **Italian formal tables** celebrate warmth and abundance. Hand-painted ceramics, vibrant tablecloths, and generous floral arrangements create a lively atmosphere. Cappuccino

glasses and grappa carafes may appear alongside standard wine glasses, honoring Italy's diverse beverage culture.

- **Scandinavian styles prioritize natural materials**—birch, linen, and simple ceramics—paired with understated elegance. Minimalist designs dominate, with monochromatic palettes and functional aesthetics that echo the region's design philosophy.

Evolution of Formal Table Presentation in the Modern Era

The twentieth and twenty-first centuries have witnessed a nuanced evolution of formal table presentation, balancing heritage with contemporary values. The post-World War II era, influenced by modernism, saw a simplification of settings: fewer pieces of cutlery, streamlined glassware, and a move toward functionalism. However, the late twentieth century experienced a revival of formal dining, driven by the luxury hospitality industry, wedding traditions, and a renewed appreciation for etiquette. Today, formal settings often blend classical elements with modern design. Sleek, minimalist crystal glassware complements traditional silver flatware; geometric centerpieces replace elaborate floral displays. Sustainability has reshaped practices: reusable linens, biodegradable dishware, and locally sourced flowers reflect eco-conscious values without sacrificing elegance. Technology, too, has discreetly entered the scene: digital menus accessed via QR codes, smart tableware that maintains temperature, and LED candle alternatives preserve aesthetics while enhancing convenience.

Contemporary Applications and Adaptations

In contemporary Europe, formal table presentation adapts to diverse contexts. State banquets and diplomatic dinners adhere strictly to tradition, ensuring every detail—from the placement of national flags on tablecloths to the sequence of toasts—aligns with protocol. Weddings often incorporate personalized elements, such as monogrammed napkins, heirloom china, or floral arrangements in the couple's chosen color palette, while maintaining the core principles of symmetry and order. Corporate galas and award ceremonies merge formality with branding, using tableware in corporate colors or featuring centerpieces that reflect the event's theme.

Fine-dining restaurants train staff to execute flawless table settings, viewing them as extensions of the culinary experience. Even in domestic settings, hosts increasingly embrace formal elements. Online platforms, etiquette consultants, and social media tutorials democratize access to these traditions, enabling individuals to create memorable, refined dining experiences at home.

Asian Influences and Minimalist Folding Styles

Asian minimalist folding is more than a set of techniques; it is a way of seeing and being in the world. Rooted in ancient philosophies and traditions, it offers a path to simplicity, mindfulness, and connection with the material and the spiritual. In an age of excess and rapid consumption, the principles of Asian minimalist folding provide a counterbalance, reminding us of the beauty that can be found in the essential. Whether through the precise creases of origami, the functional elegance of bojagi, or the thoughtful use of space in design, Asian minimalist folding invites us to slow down, to pay attention, and to appreciate the transformative power of the fold. It is a practice that transcends time and culture, offering timeless lessons in harmony, balance, and the art of living with less.

Philosophical Foundations: Taoism, Zen, and Wabi-Sabi

At the heart of Asian minimalist folding lies a philosophical framework shaped by Taoism, Zen Buddhism, and the aesthetic of wabi-sabi. Taoism, with its emphasis on the natural flow of life and the principle of wu wei (effortless action), encourages an approach to folding that is intuitive and in harmony with the material. Zen Buddhism, on the other hand, promotes mindfulness and presence, turning the act of folding into a meditative practice. The aesthetic of wabi-sabi, which finds beauty in imperfection, transience, and simplicity, informs the acceptance of asymmetry and the organic in folded forms.

These philosophies manifest in folding through the use of minimal creases, the celebration of the material's inherent qualities, and the avoidance of excess. Each fold is deliberate, yet unforced, allowing the material to speak for itself. This approach is evident in traditional Japanese origami, where a single sheet of paper is transformed into intricate forms with a limited number of folds, and in the Korean art of bojagi, where fabric is wrapped and tied with simple yet elegant techniques.

Japanese Origami: Precision and Simplicity

Japanese origami is perhaps the most iconic example of minimalist folding. Rooted in centuries-old traditions, origami embodies the principles of precision, simplicity, and resourcefulness. The use of a single square sheet of paper, without cutting or gluing, reflects a commitment to minimalism and self-constraint. Traditional models like the crane, the lotus, and the lily are achieved through a series of folds that are both economical and expressive. The minimalist ethos of origami extends beyond technique to its cultural significance. Origami is often used in ceremonial contexts, such as the folding of senbazuru (a thousand cranes) as a symbol of peace and healing. The act of folding becomes a ritual, a way to convey respect, intention, and care. Modern origami artists like Akira Yoshizawa and Sadako Sasaki have elevated the art form, demonstrating how simplicity can give rise to profound beauty and emotional depth.

Chinese and Korean Wrapping Traditions: Functionality and Elegance

In China and Korea, minimalist folding is deeply intertwined with practical traditions of wrapping and storing. The Chinese art of zhezhi (paper folding) shares similarities with origami but is often more utilitarian, used for creating containers, envelopes, and decorative items. The emphasis is on functionality, with forms that are simple yet ingenious in their design.

Fig: Stylish Napkin Folding Techniques to Impress

Korean bojagi, traditional wrapping cloths, exemplify minimalist folding in textiles. Made from scraps of fabric, bojagi are used to wrap gifts, store items, and cover food. The folding techniques are straightforward, relying on tying and draping rather than complex creases. The beauty of bojagi lies in its improvisational nature, its use of recycled materials, and its ability to transform the ordinary into something special. The minimalist approach of bojagi reflects the Korean values of resourcefulness, humility, and respect for tradition.

Materiality and Form: The Beauty of the Essential

Asian minimalist folding places a strong emphasis on materiality, allowing the inherent qualities of paper, fabric, or other mediums to guide the form. In origami, the choice of paper—its thickness, texture, and color—influences the final result. Washi, traditional Japanese paper, is often preferred for its strength and flexibility, enabling intricate folds while maintaining a delicate appearance. In bojagi, the use of patchwork fabrics highlights the beauty of imperfection and the passage of time. The folds are gentle, allowing the fabric to drape naturally, creating soft, organic shapes. This approach contrasts with Western folding traditions, which often prioritize sharpness and structure. Asian minimalist folding celebrates the material's life, its history, and its potential for transformation.

Spatial Awareness and Negative Space

A key aspect of Asian minimalist folding is the thoughtful use of space, both positive and negative. In origami, the folds create not only the form but also the empty spaces within and around it. These voids are as important as the solid shapes, contributing to the overall balance and harmony of the piece.

The concept of ma, a Japanese term referring to the interval or gap between structural parts, is central to this approach. Ma is not merely the absence of form but a presence in itself, a space that enhances the relationship between the folded object and its surroundings. This awareness of space is also evident in bojagi, where the act of wrapping creates a dialogue between the object and the cloth. The folds are arranged to complement the shape of the item, ensuring that the wrapping is neither too tight nor too loose. The result is a sense of unity and proportion, where the folded form exists in harmony with its environment.

Global Influence and Contemporary Applications

The minimalist folding styles of Asia have had a profound influence on global design, art, and architecture. In the West, the principles of simplicity, functionality, and material respect have been embraced by designers and artists seeking to counter the excesses of consumer culture. Issey Miyake's pleated garments, for example, draw on traditional Japanese folding techniques to create clothing that is both innovative and timeless.

In architecture, the concept of ma has inspired the design of spaces that prioritize openness, flexibility, and the interplay of light and shadow. The work of architects like Tadao Ando and Kengo Kuma reflects the minimalist ethos, with structures that are at once bold and restrained, complex and simple.

Middle Eastern and Mediterranean Folding Practices

Middle Eastern and Mediterranean folding practices reveal how a simple action can embody complex cultural meanings. Through garments, domestic textiles, and ceremonial cloths, folding served as a practical solution and a symbolic gesture that expressed identity, hospitality, and spirituality. Shaped by history, trade, and belief, these practices demonstrate the power of folded forms to connect material culture with social values. Understanding these traditions enriches appreciation of folding not only as a technical skill but as a living expression of regional heritage and cultural continuity.

Historical Foundations and Material Culture

The origins of folding practices in the Middle East and Mediterranean can be traced to early civilizations such as Mesopotamia, ancient Egypt, Phoenicia, Greece, and Rome. These societies developed advanced textile production using linen, wool, and later silk, materials that responded well to folding and shaping. Folding techniques emerged from everyday needs, including storage, transport, and display of cloth, as well as from ceremonial and religious requirements. The availability of fine textiles encouraged careful handling, and folding became a method of preserving fabric quality while enhancing visual presentation.

Folding in Garments and Dress Traditions

Garment folding was central to Middle Eastern and Mediterranean dress traditions, particularly in the draping and layering of clothing. Tunics, robes, and togas relied on carefully arranged folds to create structure and movement without extensive tailoring. These folds conveyed social status and regional identity, as the depth, direction, and arrangement of fabric varied according to custom and occasion. Folding techniques allowed garments to adapt to different body types and climates, demonstrating a practical intelligence that balanced comfort with elegance. The prominence of folded drapery in classical art reflects the cultural importance of these practices.

Domestic Textiles and Table Presentation

In domestic life, folding practices were essential to the organization and presentation of household textiles. Table linens, napkins, and serving cloths were folded to maintain cleanliness and demonstrate hospitality.

Fig: Elevate your Thanksgiving Table with Luxurious Designer Napkins

In Mediterranean societies, shared meals were central to social life, and the careful folding of table textiles signaled respect for guests.

Middle Eastern households similarly emphasized order and refinement in dining spaces, where folded cloths contributed to an atmosphere of generosity and care. These practices reinforced social bonds and conveyed cultural values through everyday actions.

Religious and Ceremonial Folding Practices

Religious traditions played a significant role in shaping folding practices across the Middle East and Mediterranean. In Jewish, Christian, and Islamic contexts, textiles used in worship and ritual were folded according to prescribed customs that emphasized purity and reverence. Prayer cloths, altar linens, and ceremonial coverings were handled with precision, reflecting theological concepts of order and sanctity. Folding in these settings was a disciplined act, transforming cloth into a medium of spiritual expression. The continuity of such practices highlights the enduring relationship between folding, ritual, and belief.

Influence of Trade and Cultural Exchange

The Middle East and Mediterranean regions were interconnected through extensive trade networks that facilitated the exchange of materials, techniques, and ideas. Merchants traveling along maritime and overland routes introduced new fabrics and folding styles, leading to hybrid practices that blended local traditions with foreign influences. Silk from Asia, for example, encouraged more elaborate folding techniques due to its fluidity and sheen. This constant exchange enriched regional folding practices, making them dynamic and adaptive rather than static or isolated.

Hospitality and Social Etiquette

Hospitality has long been a defining feature of Middle Eastern and Mediterranean cultures, and folding practices played a visible role in expressing this value. The presentation of folded textiles in guest spaces conveyed attentiveness and respect, signaling the host's commitment to comfort and generosity. In both regions, the careful arrangement of cushions, cloths, and table linens created inviting environments that fostered social interaction. Folding thus became a language of hospitality, communicating cultural ideals without the need for words.

Continuity and Adaptation in Modern Contexts

Despite modernization and changing lifestyles, traditional folding practices in the Middle East and Mediterranean continue to influence contemporary design and domestic routines. Modern table settings, textile crafts, and fashion designs often draw inspiration from historical folding techniques, adapting them to new materials and contexts. This continuity reflects a deep cultural appreciation for the discipline and beauty of folded forms. While methods have evolved, the underlying principles of order, balance, and respect remain central.

African Symbolic and Decorative Approaches

African symbolic and decorative approaches represent a profound synthesis of aesthetics, spirituality, and communication. Rooted in millennia of cultural practice, they encode identity, history, and cosmology through colour, pattern, and form. From the woven kente cloth to the carved masks and beaded jewellery, each decorative element serves as a testament to Africa's rich cultural tapestry. As these traditions continue to evolve—adapting to modern contexts while preserving core meanings—they affirm the continent's enduring creative genius. Understanding African decorative approaches is not merely an appreciation of beauty; it is an engagement with a living, dynamic language that speaks across generations and continents. In a globalised world, these traditions remind us that decoration, when imbued with meaning, transcends ornamentation to become a vital expression of humanity.

Historical Foundations and Cultural Significance

African decorative practices have ancient roots, with archaeological evidence—such as the rock paintings of the Sahara (e.g., Tassili n'Ajjer in Algeria) dating back 10,000 years—demonstrating early symbolic expression. These paintings depicted animals, human figures, and abstract symbols, likely tied to spiritual beliefs and environmental observations.

As agrarian societies developed, decorative arts became interwoven with ritual, governance, and daily life. For many African cultures, art is inseparable from spirituality. Masks, for instance, are not mere objects but vessels for ancestral spirits, invoked during ceremonies to mediate between the human and divine realms. The Baule people of Côte d'Ivoire craft serene, idealized masks representing beauty and harmony, while the Fang of Gabon create angular, abstracted forms used in initiation rites. Textile arts also hold ancient significance.

The bogolan (mud cloth) of the Malinké in Mali, dating to the 12th century, uses fermented mud as a resist-dye technique to imprint symbolic patterns onto cotton. These patterns narrate histories, mark social status, or convey moral lessons. Similarly, the kente cloth of the Ashanti and Ewe peoples of Ghana, woven with intricate strips of silk and cotton, encodes proverbs, historical events, and philosophical concepts through colour and geometric design. Thus, decoration in Africa is never decorative alone—it is a language, a record, and a spiritual conduit.

Symbolism in African Art and Decoration

Symbolism permeates every facet of African decorative expression. Colours, shapes, patterns, and materials carry specific meanings, often codified within cultural frameworks. In textile arts, colour symbolism is particularly pronounced. Among the Ashanti, red signifies political power and spiritual energy, blue represents peace and harmony, yellow denotes wealth and fertility, and green symbolises renewal. The arrangement of these colours in kente cloth follows strict protocols; a specific sequence might commemorate a historical event or honour a chief. Beadwork—practised extensively by the Maasai, Zulu, and Xhosa—relies on colour, shape, and placement to communicate complex social messages. For the Maasai, red beads signify bravery, blue represents the sky and water, white denotes purity, and green symbolises health and grass. A beaded necklaces colour combination might indicate the wearer's marital status, age grade, or clan affiliation. Similarly, the Zulu use beadwork to convey romantic messages: a necklace with predominantly white beads signals availability for marriage, while black and white patterns may express mourning.

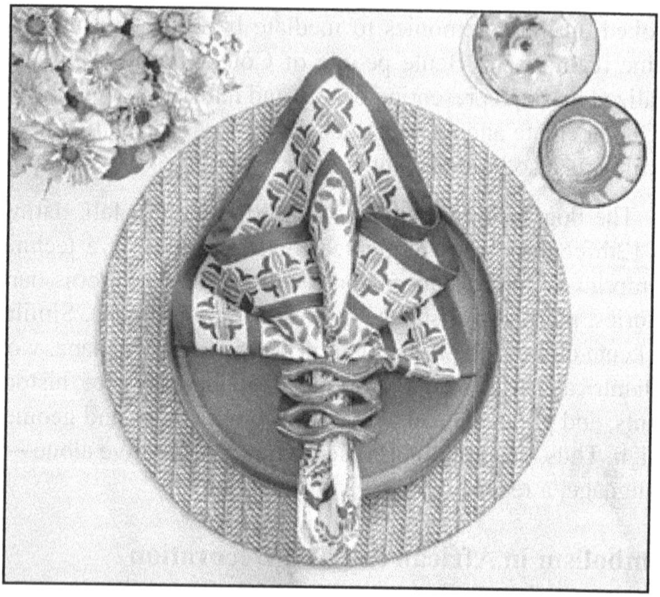

Fig: Oasis Napkin in Ivory, Green and Brown

In carved art, symbolism manifests through form. Yoruba mramram masks feature exaggerated eyes and mouths to represent specific deities or ancestors. The Dogon of Mali incorporate geometric patterns—triangles, squares, and circles—into their architecture and textiles to depict their cosmology, including the creation myth involving the Nommo twins. Even body adornment, such as scarification (mramadan) among the Nuer of South Sudan, uses patterned cuts to mark rites of passage, clan identity, or beauty standards.

Materials and Techniques

African decorative practices are deeply rooted in the continent's natural environment, utilizing locally sourced, sustainable materials. Textiles are woven from cotton, raffia palm fibre, bark cloth, or silk. The adire (tie-dye) technique of the Yoruba involves applying wax or starch to fabric before dying, creating intricate resist patterns. Bogolan employs fermented mud as a dye resist, producing earthy, symbolic designs:

- **Beadwork utilises seeds,** shells, stones, glass, or porcelain. The Maasai traditionally used natural materials like ostrich eggshell beads, though modern glass beads are now common. Techniques vary: some groups string beads by hand, while others weave them into fabric or leather.
- **Carving employs wood,** ivory, bone, stone, or metal. West African bronze casting—exemplified by the Benin Bronzes—uses the lost-wax (cire perdue) method, where a wax model is encased in clay, melted, and replaced with molten metal. Stone sculptors, such as the Shona of Zimbabwe, work with serpentine and soapstone, polishing smooth, expressive figures.
- **Painting and Body Art use Natural Pigments:** ochre, charcoal, plant dyes, and clay. The Ndebele of South Africa decorate their homes with vibrant geometric murals using natural pigments mixed with cow dung for durability. Body painting among the Himba of Namibia employs ochre mixed with butterfat to create protective patterns.

These techniques are often passed down through generations, with masters mentoring apprentices in communal settings. The process itself may involve rituals to honour the materials and the spirits believed to inhabit them.

Regional Variations

Africa's cultural diversity yields distinct decorative traditions across regions:

- **In North Africa,** Berber, Arab, and Islamic influences dominate. Geometric arabesque patterns—interlacing curves, stars, and polygons—adorn textiles, ceramics, and architecture, symbolising the infinite and the divine. Calligraphy, featuring verses from the Qur'an, is a revered decorative element.
- **West and Central Africa** are renowned for masking traditions and textile arts. The Bamileke of Cameroon craft ornate masks with polychrome finishes, while the Dan of Liberia and Côte d'Ivoire create sleek, minimalist masks embodying serenity. Textile traditions include kente cloth, bogolan, and mbramadan (raffia applique) of the Ashanti.
- **East Africa** excels in beadwork and leatherwork. The Maasai, Samburu, and Pokot use colourful beaded jewellery to convey social messages. Gourd painting and leatherwork are also prominent. In Ethiopia, sufian embroidery on garments features intricate cross-stitch patterns unique to each ethnic group.
- **Southern Africa** features the iconic Ndebele house painting, with bold geometric murals in red, white, and black symbolising the homeowner's identity. Zulu and Xhosa beadwork conveys intricate social codes, while the Shona stone sculptures reflect spiritual and ancestral themes.

Contemporary Adaptations and Global Influence

The 20th and 21st centuries have witnessed African decorative traditions evolving in response to globalisation, modernisation, and diaspora communities. Contemporary African artists blend traditional motifs with modern materials and techniques. Designers such as Duro Olowu (Nigeria) incorporate kente and adire patterns into international fashion collections. Architect David Adjaye integrates African geometric forms into modern buildings worldwide. Digital technology has transformed decorative practices. Artists use computer-aided design (CAD) to recreate traditional patterns, while social media platforms enable global dissemination of African art. Initiatives like the African Digital Art platform showcase contemporary works that reinterpret ancestral symbols. However, globalisation has also raised concerns about cultural appropriation.

Non-African designers occasionally adopt African motifs without context or credit, diluting their significance. Efforts by organisations such as Fair-Trade Africa support artisan cooperatives, ensuring ethical production and fair compensation. African diaspora communities have preserved and adapted traditions. In the Caribbean, Adinkra symbols from Ghana appear in jewellery and textiles; in Brazil, Afro-Brazilian religions incorporate Yoruba orisha iconography in altars and clothing.

American and Contemporary Fusion Styles

American and contemporary fusion styles exemplify the dynamic interplay between tradition and modernity, locality and globalization, craftsmanship and technology. Through their distinctive yet interconnected characteristics, these styles enrich creative expression across disciplines and contexts. By studying fusion styles in academic and vocational frameworks, learners gain a nuanced understanding of cultural dialogue, design principles, and innovation strategies. This foundation prepares them to contribute thoughtfully and creatively to the evolving landscape of contemporary design and culture.

Historical Context of American Style

The American style has been shaped by a confluence of cultural, historical, and geographic factors. Rooted in colonial influences and Native American traditions, it evolved through periods such as the Victorian era, Arts and Crafts movement, and mid-century modern design. Characterized by practicality, comfort, and eclecticism, traditional American style incorporates natural materials, handcrafted details, and a sense of warmth. Academic studies often examine how American design reflects social values like individualism, adaptability, and resourcefulness, serving as a foundation for fusion innovations.

Fig: Napkin Folds to Elevate the Look of your Tablescape

Emergence of Contemporary Fusion Styles

Contemporary fusion styles emerged in response to rapid cultural exchange, technological advancement, and shifting consumer tastes in the late 20th and early 21st centuries. This approach deliberately combines elements from different time periods, cultures, and materials to produce fresh and unexpected outcomes. Fusion is not limited to aesthetics; it also embodies a philosophy of inclusivity and experimentation. In educational programs, contemporary fusion is studied as a methodology that challenges traditional boundaries and encourages cross-disciplinary collaboration.

Characteristics of American Fusion Style

American fusion style synthesizes traditional American motifs with influences from diverse cultures such as Latin American, Asian, and African design. This style often features bold colors, mixed textures, and layered patterns that create a rich sensory experience. Materials like reclaimed wood, industrial metals, and artisanal textiles are combined to balance rustic charm with urban sophistication. The fusion reflects America's multicultural landscape and celebrates diversity while maintaining a distinct sense of place.

Characteristics of Contemporary Fusion Style

Contemporary fusion style prioritizes minimalism, clean lines, and innovative materials while integrating cultural or historical references. It frequently juxtaposes sleek, modern forms with handcrafted or organic details to produce tension and harmony. Use of technology, sustainable materials, and adaptive design principles are common. This style is versatile, adaptable to various contexts from residential spaces to commercial environments. Academically, contemporary fusion is analyzed as a response to globalization and digital connectivity, representing the zeitgeist of the 21st century.

Application of Fusion Styles across Disciplines

Fusion styles extend beyond design aesthetics to influence culinary arts, fashion, music, and media. In interior design, fusion may merge mid-century modern furniture with ethnic textiles. Fashion combines streetwear with traditional craft techniques. Culinary fusion blends diverse ingredients and cooking methods to create new flavors. In media, fusion manifests in hybrid genres and mixed cultural narratives. Academic studies explore these applications to understand how fusion fosters creativity and cultural dialogue in multiple domains.

Table: Comparative of American and Contemporary Fusion Styles

Aspect	American Fusion Style	Contemporary Fusion Style
Cultural Influences	Multicultural American heritage	Global and multicultural influences
Design Aesthetic	Eclectic, warm, layered	Minimalist, sleek, innovative
Material Use	Natural, reclaimed, artisanal	Technological, sustainable, experimental
Color Palette	Bold, earthy, rich	Neutral, monochrome, accent-focused
Emotional Impact	Comfort, nostalgia, vibrancy	Sophistication, innovation, clarity
Typical Applications	Residential, hospitality	Commercial, residential, public spaces
Philosophical Approach	Celebrates diversity and place	Challenges boundaries and norms
Craftsmanship	Emphasis on handmade and heritage	Mix of handcrafted and digital fabrication
Adaptability	Grounded in tradition with modern twists	Fluid, trend-responsive, versatile
Educational Focus	Cultural identity and materiality	Globalization and technology integration

Principles Guiding Fusion Style Design

Successful fusion style design adheres to several guiding principles that ensure cohesion, relevance, and creativity:

- Respect for cultural origins and context.
- Balance between contrasting elements.
- Thoughtful selection of materials and textures.
- Integration of technology and craftsmanship.
- Attention to proportion and spatial harmony.
- Emphasis on storytelling and meaning.
- Flexibility to adapt to different environments.
- Sustainability and ethical sourcing.
- User experience and functionality.
- Continuous innovation and experimentation.

Challenges and Opportunities in Fusion Design

While fusion styles offer rich creative potential, they also pose challenges such as cultural appropriation, stylistic incoherence, and technical complexity. Academic discourse encourages critical examination of how fusion navigates authenticity and innovation. At the same time, fusion styles provide opportunities to bridge cultures, introduce novel aesthetics, and foster inclusive design practices. Educators emphasize reflective practice and research to empower designers to engage fusion styles responsibly and effectively.

Educational Relevance and Professional Impact

In design education, American and contemporary fusion styles are studied to cultivate cultural literacy, creative problem-solving, and interdisciplinary collaboration. Students learn to analyze historical precedents, current trends, and future directions. Vocational training includes hands-on projects that simulate real-world fusion design challenges. Professionally, mastery of fusion styles enhances marketability and relevance in a globalized creative economy. Designers equipped with fusion competencies can respond to diverse client needs and push innovation boundaries.

Standards in Hotels, Restaurants, and Catering

Importance of Standards in the Hospitality Industry

Standards are crucial because hospitality services are largely intangible and experiential, making quality difficult to measure without clear benchmarks. In hotels, restaurants, and catering services, standards define acceptable levels of service performance, hygiene, safety, and ethical conduct.

They protect guests from health risks, ensure fair treatment of employees, and create operational discipline within organizations. From a business perspective, standards reduce variability, minimize errors, and support cost control, while from a customer perspective, they build trust and confidence. A guest who visits a hotel or restaurant expects a predictable level of comfort, cleanliness, and service efficiency, and standards make such predictability possible.

Regulatory and Legal Standards

Regulatory standards are enforced by government and statutory bodies to safeguard public health, safety, and consumer rights. In hospitality, these standards typically cover food safety, fire safety, building codes, labor laws, and environmental protection. Restaurants and catering operations must comply with food hygiene regulations, proper storage and handling of ingredients, and regular health inspections.

Hotels are required to meet standards related to occupancy limits, emergency exits, fire detection systems, and accessibility. Compliance with legal standards is non-negotiable, as violations can lead to fines, closures, or reputational damage. These regulations also provide a baseline upon which organizations can build higher voluntary standards.

Service Quality Standards and Guest Experience

Service quality standards focus on how guests are treated and how services are delivered. In hotels, this includes front-desk efficiency, housekeeping consistency, room amenities, and responsiveness to guest requests. In restaurants, service standards govern order accuracy, serving time, staff behavior, and complaint handling.

Catering services rely heavily on standards related to event coordination, punctuality, food presentation, and adaptability to client needs. These standards help translate abstract concepts such as courtesy, professionalism, and hospitality into observable and measurable practices. By setting clear expectations for staff behavior and service delivery, organizations can create memorable and positive guest experiences.

Food Safety and Hygiene Standards

Food safety and hygiene standards are among the most critical in restaurants and catering, but they are equally important in hotel kitchens and banqueting operations.

Fig: Health Safety and Hygiene Rules in the Kitchen

These standards cover personal hygiene of staff, cleanliness of equipment, pest control, temperature control, and prevention of cross-contamination. International frameworks such as Hazard

Analysis and Critical Control Points (HACCP) provide systematic approaches to identifying and managing food safety risks. Adherence to these standards not only prevents foodborne illnesses but also reinforces customer confidence. In a highly competitive market, a reputation for safe and hygienic food practices can be a significant competitive advantage.

Operational and Facility Standards

Operational standards address the efficiency and consistency of daily activities, while facility standards focus on the physical environment. In hotels, operational standards may include check-in and check-out procedures, maintenance schedules, and inventory management. Facility standards cover room size, lighting, ventilation, cleanliness, and accessibility features.

Restaurants and catering facilities must meet standards related to kitchen layout, equipment maintenance, waste disposal, and storage capacity. These standards ensure that operations run smoothly and that physical spaces support both employee productivity and guest comfort. Well-maintained facilities also signal professionalism and care, influencing overall customer perception.

Human Resource and Training Standards

Employees are central to hospitality service delivery, making human resource standards especially important. These standards relate to recruitment, training, performance evaluation, workplace safety, and ethical conduct. Hotels, restaurants, and catering companies often establish training standards to ensure that employees possess the necessary technical skills, product knowledge, and interpersonal abilities. Consistent training aligned with service standards helps reduce service gaps and enhances teamwork. Additionally, fair labor practices and clear codes of conduct contribute to employee satisfaction and retention, which in turn positively affect service quality and organizational reputation.

Sustainability and Ethical Standards

In recent years, sustainability and ethics have become integral to hospitality standards. Hotels are increasingly adopting standards related to energy efficiency, water conservation, waste reduction, and responsible sourcing. Restaurants and catering services focus on minimizing food waste, using eco-friendly packaging, and supporting local producers. Ethical standards also address issues such as fair trade, animal welfare, and community engagement. These standards respond to growing consumer awareness and demand for socially responsible businesses. By integrating sustainability into operational standards, hospitality organizations can reduce environmental impact while strengthening brand value.

Table: Standards across Hospitality Sectors

Aspect	Hotels	Restaurants	Catering Services
Primary Focus	Accommodation and guest comfort	Food and dining experience	Event-based food and service delivery
Key Safety Standards	Fire safety, building codes	Food hygiene, sanitation	Food transport and storage safety
Service Standards	Front-desk, housekeeping, concierge	Table service, order accuracy	Timeliness, customization
Facility Standards	Rooms, common areas, amenities	Dining area, kitchen layout	Temporary and mobile setups
Regulatory Oversight	Tourism and safety authorities	Health and food authorities	Health and local authorities

Elements Commonly by Hospitality Standards

- Cleanliness and hygiene of all guest-facing and operational areas.
- Safety measures for guests, staff, and property.
- Consistency in service delivery across shifts and locations.

- Professional appearance and conduct of staff.
- Quality and safety of food and beverages.
- Maintenance and upkeep of facilities and equipment.
- Clear procedures for handling complaints and feedback.
- Compliance with legal and regulatory requirements.
- Training and development of employees.
- Commitment to sustainability and ethical practices.

Challenges in Implementing and Maintaining Standards

Despite their importance, implementing and maintaining standards can be challenging. Variations in staff skill levels, high employee turnover, cost constraints, and cultural differences can affect consistency. Small restaurants and catering businesses may struggle with the financial and administrative burden of compliance, while large hotels may face difficulties in standardizing practices across multiple departments.

Continuous monitoring, regular audits, and ongoing training are necessary to overcome these challenges. Technology, such as digital checklists and automated reporting systems, is increasingly used to support standard compliance and performance evaluation.

Napkin Folding for Banquets and Formal Events

Classic European Repertoire

In the great dining rooms of Europe, especially in France and Austria from the eighteenth century onward, certain folds became canonical. The Bishop's Mitre, tall and ecclesiastical, still crowns plates in Vienna's state banquets and in the Élysée Palace on formal nights. Its sharp central peak is formed by rolling the napkin into a tube, folding it in half, then tucking and standing the two resulting points like the twin flaps of a bishop's hat. Nearby, the Fleur-de-Lis fold, with its three upright petals, pays quiet homage to the old French monarchy and appears whenever protocol demands unmistakable elegance. Perhaps the most beloved of all is the Rose, created by rolling the napkin diagonally, then coiling it tightly from one corner and gently easing the outer layers outward until a lush linen bloom rests in the water glass or lies flat on the charger.

Candle and the Fan: Light and Movement

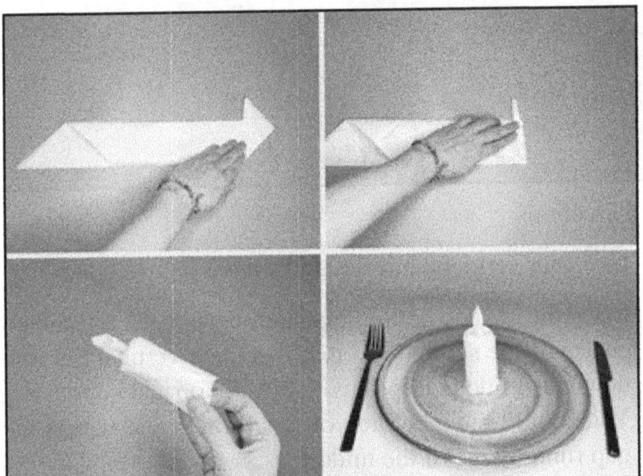

Two folds in particular seem to capture motion itself. The Candle, slender and upright in the goblet, mimics beeswax tapers and casts a soft shadow across the silver. It is achieved by folding the napkin into a narrow rectangle, rolling it tightly from the bottom, and turning back the top inch to form a convincing flame. Beside it, the Fan (sometimes called the Accordion or the Lady's Fan) rises in crisp pleats from a bread plate or stands behind the place card like an ivory screen. To create it, the napkin is folded back and forth in narrow even pleats, gathered at the base, and fanned open at the top. When a dozen such fans line a long table, the effect is one of gentle undulation, as though a breeze had just passed through the room.

Directions:

- Lay your serviette out flat and square before you.
- Bring the top and bottom right corners into the top and bottom centre of the serviette so that there is a vertical line running down the middle. Repeat with the left top and bottom corners. You should have before you a tall rectangle with a vertical gap running down the middle.
- Place your finger in the very centre of the serviette where the two flaps meet each other. With your other hand take the top right flap and pull it to the right, going past the top of the rectangle so that it makes a triangle shape a little like the flap on a collar. Repeat with the top point on the left side. It now looks like the collar on a shirt with rather a wide neck at the back.
- The next step involves making a mirror image of the last step at the bottom, if it's easier then carefully turn your serviette around do that the collar is at the bottom upside down and then just repeat the previous step at the top as before. Once all

four flaps are completed you will have before you a rectangle that has 4 triangles radiating out from the centre a little like a windmill.

- Starting at the bottom upside-down collar part, take the bottom left and right edges and roll them as tightly as you can up to the centre of the serviette. (I set a glass on mine to keep it there while I did the next step).
- Turn the entire serviette around to the that roll is at the top and the collar bit of the top half is now upside down at the bottom.
- Accordion fold the bottom bit up to the centre (it will meet the roll at the middle).
- Now gather both ends (without letting it all unfold) and the roll should be in the middle.
- Place it into an empty water or wine glass and fan out the accordion fan bits at each side. So easy!

Pocket and the Envelope: Practical Grandeur

Many modern banquets favor folds that serve a purpose beyond ornament. The Single Pocket, formed by folding the napkin into quarters and then turning down one corner to create a diagonal cuff, receives the menu card or a small spray of flowers. Its cousin, the Triple Pocket, divides the napkin into three ascending sleeves, each capable of holding flatware or a printed program. The Envelope fold, crisp and architectural, seals the silverware completely inside a square packet tied with a ribbon or secured by a final tuck. At state dinners, where security once required that no knife lie openly on the table until the guest is seated, the Envelope was both beautiful and discreet.

National Signatures and Diplomatic Language

Napkin folding has long been a form of soft diplomacy. At a banquet hosted by the Japanese Imperial Household, guests may find the Tsuru (crane) fold resting on their plate, its long neck and beak formed by a series of precise valley and mountain creases. The crane, symbol of happiness and longevity, wordlessly conveys the host's wishes.

In Thailand, the Lotus Bud appears at royal dinners, its pointed tip rising from layered petals created by rolling and inverting the corners. Scandinavian tables favor the understated but ingenious Herringbone or the standing Sail, both of which lie almost flat yet catch light along their sharp ridges. An ambassador quickly learns to read these folds the way courtiers once read the height of a candlestick or the placement of a salt cellar.

Buffet Table Spectacle

Banquets serving hundreds of guests often move the drama from the individual place setting to the central buffet. Here towers reign. The towering Pineapple, built from six or eight napkins stacked and pleated upward, can rise nearly two feet and serves as an edible-looking centerpiece that is, in fact, pure linen. The Waterfall, created by draping dozens of napkins in staggered overlapping rows down the front of the table, gives the illusion of cascading fabric. At wedding receptions, swans with arched necks face each other across the sweet table, their bodies formed by rolling and shaping two stiffly starched napkins around crumpled paper cores to hold the curve.

Starch, Iron, and Discipline

None of these creations would be possible without rigorous preparation in the linen room. Banquet napkins destined for elaborate folds are typically 20 to 24 inches square, woven from Irish linen or a cotton-linen blend heavy enough to hold a crease. They are washed without fabric softener, starched lightly (never heavily enough to crack), and pressed while still damp. The most demanding folds, such as the Rosebud or the Mitre, are executed on a wide padded table by two or three stewards working in silence.

A steam iron hovers nearby to sharpen any softening edge. In the finest houses, each finished napkin is slipped into a labeled cardboard ring to protect the shape until the moment it is placed on the table.

Moment of Unfolding

The life of a formal napkin fold is brief and poignant. After the guest is seated, the sculpture is undone with a single gesture. The Bishop's Mitre collapses into a flat square, the Fan closes like a book, the Candle is shaken out and laid across the lap. This deliberate destruction is part of the ritual; the napkin must sacrifice its beauty to serve its purpose.

Yet in that brief existence (sometimes no more than ten minutes) it has done its work: it has signaled that the evening is special, that attention has been paid, and that every detail, even one as small as a folded square of linen, matters. In the end, the art of napkin folding for banquets is less about the napkin itself than about the intention behind every crease. It is a language of hospitality spoken without words, a promise that the hours to come will be ordered, gracious, and worthy of memory. When the last candle gutters and the final toast is drunk, the linen is gathered, laundered, and pressed flat again, waiting for the next occasion when it will rise once more into silent, impeccable sculpture.

Branding and Identity through Table Presentation

Branding and identity through table presentation represent a sophisticated interplay between design, culture, and experience. By transforming tables into expressive platforms, brands can communicate their values, aspirations, and distinctiveness in subtle yet impactful ways. Effective table presentation goes beyond decoration, functioning as a strategic tool that shapes perception, evokes emotion, and reinforces recognition. In an increasingly experience-driven marketplace, the ability to articulate brand identity through thoughtful table presentation offers a powerful means of connection, allowing guests to engage with a brand not only visually but also emotionally and culturally.

Concept of Branding in Table Presentation

Branding refers to the deliberate creation of a recognizable and consistent identity that distinguishes one organization or experience from another. In the context of table presentation, branding manifests through repeated visual and sensory elements that align with a broader brand narrative.

Fig: Creative Napkin Folding for Modern Summer Table

The selection of tableware, linens, centerpieces, and spatial layout contributes to a unified impression that reinforces brand recognition. When these elements are thoughtfully coordinated, the table becomes an extension of the brand itself, reflecting its ethos, target audience, and market positioning. This approach ensures that guests encounter the brand not only through logos or advertisements but also through immersive, tangible experiences.

Visual Identity and Aesthetic Coherence

Visual identity plays a central role in translating brand values into table presentation. Color palettes, textures, and forms must work in harmony to create aesthetic coherence. A minimalist brand may favor clean lines, neutral tones, and uncluttered arrangements, while a heritage-inspired brand might emphasize rich fabrics, ornate details, and traditional motifs. Consistency across tables within a venue reinforces the brand image, creating a sense of reliability and professionalism. At the same time, visual identity should be adaptable enough to accommodate seasonal themes or special occasions without compromising the core brand message.

Materials, Craftsmanship, and Brand Values

The materials used in table presentation communicate important aspects of brand identity. High-quality porcelain, crystal glassware, and finely woven linens often signal luxury, refinement, and attention to detail. In contrast, handcrafted ceramics, natural wood, or recycled materials may express authenticity, sustainability, and a connection to nature.

Craftsmanship also plays a symbolic role, as well-made items reflect care, expertise, and respect for tradition or innovation. Through material choices, brands can subtly convey ethical commitments, cultural influences, and quality standards, allowing guests to form deeper emotional connections with the dining experience.

Cultural Influences and Identity Expression

Table presentation is deeply influenced by cultural context, making it a powerful medium for expressing identity in a globalized world. Traditional table settings can celebrate regional heritage, showcasing distinctive patterns, utensils, and arrangements that reflect local customs. At the same time, contemporary brands often blend cultural references to create hybrid identities that appeal to diverse audiences. This fusion can communicate openness, creativity, and global awareness. However, authenticity remains crucial, as thoughtful integration of cultural elements demonstrates respect and understanding, while superficial imitation risks diluting the brand's credibility.

Role of Consistency in Brand Recognition

Consistency is fundamental to effective branding through table presentation. Repeated exposure to recognizable visual cues helps guests associate certain styles and arrangements with a specific brand. This consistency extends beyond individual elements to encompass overall atmosphere, service style, and spatial organization. When guests encounter a familiar table presentation across different locations or events, it strengthens brand recall and trust. At the same time, consistency does not imply rigidity, as successful brands balance uniformity with subtle variation to maintain interest and relevance while preserving their identity.

Emotional Engagement and Guest Perception

Branding through table presentation is closely linked to emotional engagement, as visual and tactile elements influence how guests feel about an experience. A thoughtfully designed table can evoke comfort, excitement, elegance, or nostalgia, shaping the emotional tone of a meal or event.

These emotional responses contribute to how guests perceive the brand as a whole, often leaving a lasting impression that extends beyond the immediate experience. By aligning table presentation with desired emotional outcomes, brands can create memorable moments that foster loyalty and positive word-of-mouth.

Dress Up Your Place Settings with Fancy Napkin Folds

Meticulous attention to detail is the key to giving your wedding celebration an air of elegance that your guests won't soon forget.

- The place settings, the centerpieces, even the way the napkins are folded make an impression.
- At Arrowhead Golf Club, when you plan your special day, you may select from a myriad of fancy napkin folds to give your event an extra infusion of panache and polish.
- "During our wedding tastings, we do display each napkin fold," said Olivia Englert, Arrowhead sales and wedding coordinator. Couples may choose from the display.

- "These are perfect for weddings if they are doing thank-you cards, or menu cards," Olivia said. Even items such as chocolate bar favors can be inserted in the pocket for each guest.
- Another fold popular at weddings is the spiral fold that Olivia calls simple and elegant.

- The rose fold imitates the petals of a blooming flower and are placed inside wine glasses.

- The tent fold is a standard option often used for showers and corporate events that is sometimes used at weddings. There is also an intricate fold known as the bishop's hat that is most frequently used at fundraisers.

- The tuxedo fold, or the "bath robe" as staff like to call it, is typically used for the cake table,
- The drape, or hang-over fold, extends over the edge of the table.

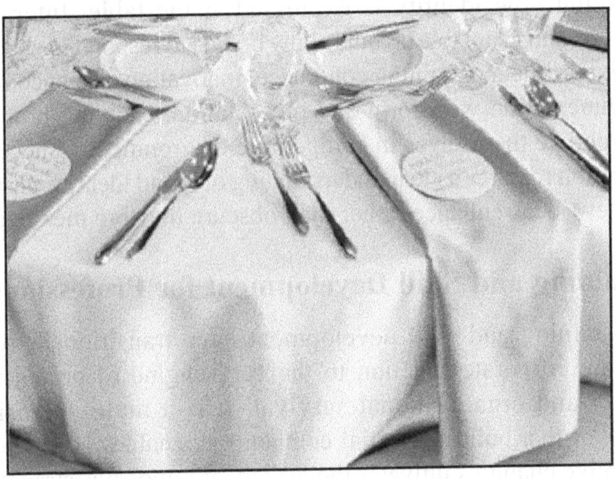

- "This option is beautiful with a colored napkin against our standard ivory linen,"
- Ivory and black linen napkins are complimentary. Couples may choose other colors for an extra charge.

Table Presentation in Corporate and Event Branding

In corporate and event contexts, table presentation serves as a strategic branding tool that reinforces organizational identity and purpose. Conferences, product launches, and formal dinners often use customized table elements to reflect corporate values and messaging. Subtle branding through color schemes, layout, and material selection can communicate professionalism, innovation, or prestige without relying on overt logos. In such settings, the table becomes part of a broader narrative that supports brand storytelling and enhances the overall impact of the event.

Innovation and Adaptation in Contemporary Branding

As consumer expectations evolve, brands increasingly seek innovative approaches to table presentation. Technological integration, modular designs, and sustainable practices are reshaping how identity is expressed at the table. Interactive elements and adaptable layouts allow brands to respond to changing contexts while maintaining coherence. Innovation also encourages experimentation with form and presentation, enabling brands to differentiate themselves in competitive environments. Successful adaptation requires a clear understanding of brand identity, ensuring that new ideas enhance rather than obscure the core message.

Training and Skill Development for Professionals

Training and skill development have transitioned from a periodic corporate function to the very engine of professional vitality and organizational survival. It is a multidimensional endeavor that builds essential competence, enables adaptation to disruptive change, cultivates the leaders of tomorrow, strengthens organizational fabric, and empowers individual agency. For the professional, it is the ongoing practice of staying relevant, engaged, and fulfilled. For the organization, it is the strategic investment that fuels innovation, loyalty, and long-term resilience. In a world where the only constant is change, the commitment to continuous learning is the ultimate professional imperative and the most reliable compass for navigating the future.

Foundations of Professional Competence and Confidence

At its core, systematic training serves to establish and fortify the foundational platform upon which professional competence is built. Initial academic education provides the theoretical framework, but it is through targeted, practical training that theory is translated into actionable skill. This process is essential for building not only proficiency but also the deep-seated confidence required to perform effectively.

When a professional undergoes comprehensive training in a new software platform, a revised regulatory framework, or an advanced technical methodology, they are doing more than acquiring information. They are internalizing processes, developing muscle memory for complex tasks, and constructing a mental model for problem-solving within their domain. This cultivated expertise directly reduces errors, increases efficiency, and enhances the quality of output. Moreover, the confidence gained from mastering new competencies empowers professionals to take on greater responsibilities, contribute ideas more freely, and navigate challenges with assuredness. This solid foundation of competence and confidence is the bedrock of individual performance and the first, non-negotiable pillar of a thriving career.

Navigating the Tides of Technological and Market Change

The most pressing driver for continuous skill development is the relentless pace of change itself. Technological advancements, particularly in fields like artificial intelligence, data analytics, automation, and digital communication, are continuously redefining job roles and industry standards. Market dynamics shift overnight due to global events, changing consumer behaviors, and new competitive threats. Professionals who stand still, relying solely on the knowledge they acquired years prior, inevitably find their skills eroding in relevance. Proactive training is the mechanism for navigating these tides.

It is the deliberate act of future-proofing one's career. For instance, a marketing professional learning data science principle, an accountant mastering blockchain applications for auditing, or a manufacturing engineer studying predictive maintenance powered by the Internet of Things, are all engaging in strategic adaptation. This forward-looking approach to development ensures that a professional remains an asset rather than becoming a legacy cost. It enables both individuals and organizations to not just react to change, but to anticipate and leverage it, turning potential disruption into strategic opportunity.

Cultivating Leadership and Strategic Soft Skills

While technical prowess is crucial, the elevation of a professional into leadership and strategic roles hinges on the parallel development of sophisticated soft skills. Training in these areas is often more nuanced but equally vital. As careers progress, the value of expertise in communication, emotional intelligence, conflict resolution, ethical decision-making, and strategic thinking increases exponentially.

Dedicated training programs in these domains are essential for moving from a specialist contributor to an effective manager, and from a manager to an inspirational leader. Such development might involve executive coaching, workshops on inclusive leadership, courses in negotiation and persuasion, or training in change management methodologies. These skills enable professionals to motivate teams, foster collaborative innovation, manage stakeholder relationships, and articulate a compelling vision. They transform raw talent and technical knowledge into influential leadership. Investing in this dimension of professional growth ensures that an organization cultivates not just skilled workers, but the empathetic, resilient, and visionary leaders necessary to guide it through complexity.

Fostering Engagement, Retention, and a Culture of Innovation

From an organizational perspective, a robust commitment to employee training and development is a powerful strategic lever with direct benefits for culture and performance. When a company invests in the growth of its people, it sends a powerful message about their value and its commitment to their long-term success. This investment is a key driver of employee engagement and job satisfaction. Professionals feel recognized, challenged, and valued, which in turn fosters loyalty and drastically reduces turnover.

The high costs associated with recruiting and onboarding new talent are mitigated by investing in the growth of existing talent. Furthermore, a culture that prioritizes learning becomes a culture of innovation. When employees are encouraged to acquire new knowledge, experiment with new ideas, and apply fresh perspectives, they become agents of improvement. Cross-functional training breaks down silos, sparking collaboration and the cross-pollination of ideas. This environment, where development is woven into the fabric of the organization, attracts ambitious talent and creates a self-reinforcing cycle of improvement, adaptability, and competitive advantage.

Personal Dimension: Autonomy, Fulfillment, and Career Agility

Beyond organizational mandates, personal agency in skill development is perhaps the most transformative aspect for the modern professional. In an era where career paths are increasingly non-linear and portfolio-based, taking ownership of one's learning journey is an act of career self-determination. This involves self-assessment, curiosity, and the initiative to seek out resources—be it online courses, professional certifications, industry conferences, or mentorship relationships. This self-directed development cultivates a growth mindset, where challenges are viewed as opportunities to learn rather than insurmountable obstacles. The process itself brings intrinsic fulfillment, combating stagnation and reigniting passion for one's work.

]Ultimately, it builds career agility—the invaluable ability to pivot, seize new opportunities, and design a professional life that is resilient, dynamic, and aligned with one's evolving aspirations. The professional who consistently learns is not at the mercy of the market; they are equipped to shape their own destiny.

Time Management and Operational Efficiency

Time management and operational efficiency are inseparable elements of effective organizational performance. By treating time as a strategic resource and embedding time management principles into operational practices, organizations can achieve higher productivity, lower costs, and improved quality. The integration of optimized processes, supportive technology, and a strong efficiency-oriented culture creates a foundation for long-term success. In an increasingly competitive and dynamic environment, organizations that master time management are better positioned to deliver value consistently and sustainably.

Relationship between Time and Efficiency in Organizations

Time and efficiency are closely interrelated, as inefficient use of time often results in operational bottlenecks, increased costs, and employee stress. In organizational settings, delays in one process frequently cascade into other areas, disrupting schedules and lowering performance standards. Effective time management allows organizations to align tasks with priorities, coordinate activities across departments, and ensure that resources are deployed at the right moment. By understanding time as a strategic asset rather than merely a scheduling concern, organizations can design processes that minimize idle periods and maximize productive output.

Principles of Effective Time Management

Effective time management is built on principles such as goal clarity, prioritization, planning, and disciplined execution. Clearly defined goals help individuals and teams focus their time on activities that contribute directly to organizational objectives. Prioritization ensures that critical tasks receive attention before less important ones, reducing the risk of missed deadlines or rushed work. Planning translates goals into actionable schedules, while disciplined execution involves adhering to plans and adjusting them when circumstances change. Together, these principles create a structured approach to managing time that supports operational efficiency across all levels of an organization.

Operational Efficiency as a Strategic Objective

Operational efficiency is not merely a technical concern but a strategic objective that influences competitiveness and profitability. Efficient operations enable organizations to deliver products and services faster, reduce operational costs, and improve customer satisfaction. Time management plays a central role in achieving this efficiency by eliminating unnecessary steps, reducing waiting times, and improving coordination. Organizations that prioritize operational efficiency are better equipped to scale their activities, adapt to market changes, and sustain long-term growth. This strategic perspective encourages continuous evaluation and refinement of time-related practices.

Process Optimization and Workflow Design

Process optimization involves analyzing existing workflows to identify delays, redundancies, and non-value-adding activities. Time management contributes to this effort by highlighting how long tasks take and where time is being lost. By redesigning workflows to ensure logical sequencing and clear accountability, organizations can significantly improve operational efficiency. Efficient workflow design reduces handoff delays, clarifies roles, and ensures that tasks progress smoothly from initiation to completion. Over time, optimized processes lead to predictable outcomes and improved organizational reliability.

There are so many things I enjoy about hosting dinner parties. The food, booze, company and table setting makes the whole experience fantastic. Yes, I said table setting. I have a ton of gorgeous linens but napkin folding has always been a challenge for me. Do you agree? I always try to hide a napkin under a plate just to get by. Luckily, I researched and practiced my napkin folding techniques for a while now to bring you this absolutely necessary article. Here are 10 ways to fold napkins for every occasion. Save this article because you're going to use this throughout the year.

Ways to Fold Napkins

I came up with easy ways to fold napkins that anyone can do. These folds and table setting is acceptable for all occasions and seasons. You can switch these up and use them for whatever occasion you want.

Summer Cookout

This fold is one of the easiest. I call it the Summer Cookout because this was the fold I did for a summer cookout once so the name stuck with me. Fold the napkin diagonally. Fold the corners up to meet the top corner. Fold the napkin back and close the corners in.

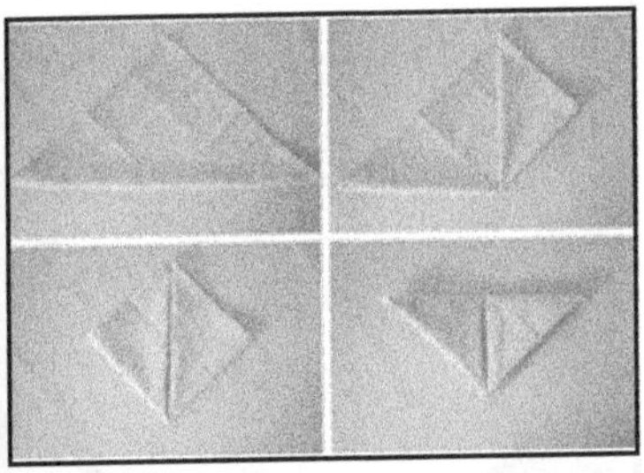

Summer Corn

This fold is harder. Fold the napkin diagonally. Fold the corners up to the top point of the triangle. Fold the corner of the flaps back down so it's a half triangle. Fold the back triangle tip down 3/4 of the way. Flip the napkin over and bring the side angles in like hugging arms. Tuck the angles into the pockets of each other. Peel the top layer down on each side like a corn husk. That's why I call it Summer Corn.

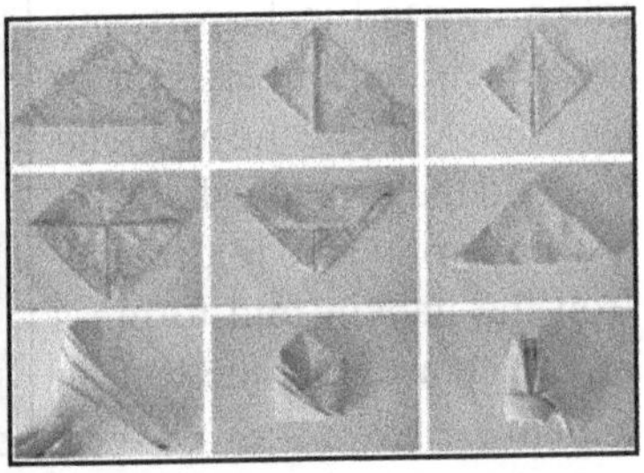

Summer Trifecta

I call this a Summer Trifecta because the napkin has three slots for a spoon, fork and knife. Fold the napkin two ways into a square. Fold the corner down evenly. Fold the second layer down and tuck the corner behind the first layer. Fold the third layer down the same way as the second layer. Fold the left side back 1/3 of the square. Then fold the right side the exact same way. You should have a slot for all three utensils.

Fall Leaf

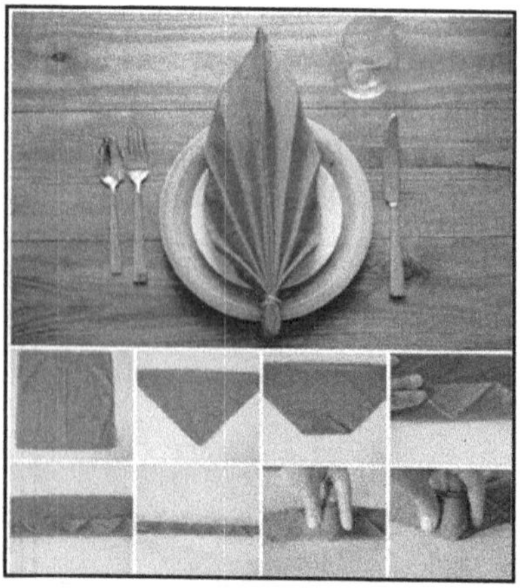

This giant fall leaf is eye catching when you use a popping color like orange. Fold the napkin diagonally. At the bottom tip of the triangle, fold a zig zag 1 inch apart until you reach the longest side of the triangle. Bend the midpoint and fold the napkin in half. Tie a string around the folded end. Iron the napkin fold to press on the crease. Lay the leaf on a plate.

Fall Thankful

Fold a napkin diagonally. Bring the middle tip into the middle of the longest side. Take the top point of the triangle and fold it down but not crossing the middle point. From the top, start rolling the napkin all the way down. Tie the roll with a string. The napkin should look like a candle.

Winter Holiday

This is a good table setting when your guests are sitting very close together. The flatware are tucked into the napkin. Start by folding the napkin in half long way. Fold one half down to meet the middle crease of the napkin. Fold the napkin in half and then in quarter leaving a middle pocket to tuck the flatware. Add Rosemary or holy berries for some holiday cheer.

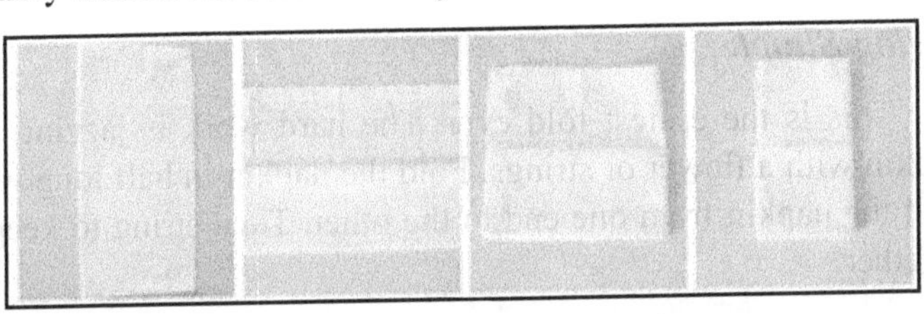

Winter Bow

Start by folding the napkin diagonally. Fold the longest side 1 inch until you reach the middle. Fold the middle point of the triangle in 2 inches. Continue folding the rest of the napkin all the way. Bend the top leg down and bottom leg up making a X. Pinch the middle and tie a glitzzy string to make the bow.

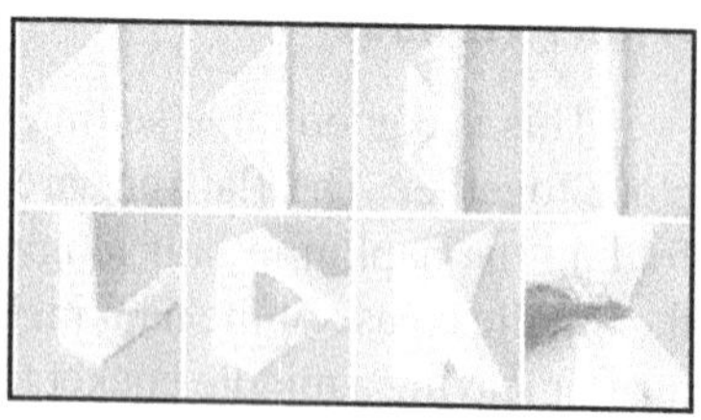

Spring Simple

This is the easiest fold ever. The hard work is jazzing the napkin with a flower or strings. Fold the napkin in half long way. Roll the napkin from one end to the other. Tie a string to keep it together.

Spring Bunny

This is the hardest fold ever but worth the effort because the napkin looks like a bunny rabbit. It's perfect for Easter. Fold the napkin in half long way and then half again. Fold the ends down from the middle to look like a house. Fold the bottom corners up diagonally. Turn the napkin around and bring the side corners in to meet in the middle. Fold the sides in and tuck the corners in between the folds. Make sure the tuck is tight. Turn the napkin around and fluff the bunny ears.

Summer Rose

This is an adorable fold for weddings and summer garden

parties. Start by folding the napkin diagonally. From the longest side, fold 1 inch again and again to 3/4 of the way. From the longest end, roll the napkin all the way over. Tuck the remaining flap into the loose flap above the roll. Fold the two flap over to cup the rose. Add a green leaf under the rose.

Role of Technology in Enhancing Time Management

Technology has become a powerful enabler of time management and operational efficiency. Digital tools such as project management systems, automated scheduling software, and real-time performance dashboards help organizations monitor time usage and coordinate activities. Automation reduces the time spent on repetitive tasks, allowing employees to focus on higher-value work. Communication technologies also accelerate information flow, minimizing delays caused by miscommunication or lack of coordination. When effectively integrated, technology supports data-driven decision-making and continuous improvement in time-related performance.

Human Factors and Organizational Culture

While systems and tools are important, human behavior and organizational culture significantly influence time management outcomes. A culture that values punctuality, accountability, and respect for deadlines reinforces efficient time use. Training programs that develop time management skills empower employees to plan their work effectively and manage competing demands.

Leadership behavior also sets the tone, as managers who model good time management practices encourage similar behavior throughout the organization. Aligning cultural values with operational goals creates an environment where efficiency is naturally sustained.

Core Practices Supporting Time Management and Efficiency

- Clear definition of goals and performance expectations.
- Systematic prioritization of tasks and activities.
- Structured planning and scheduling of work.
- Continuous monitoring of time usage and outcomes.
- Elimination of non-value-adding activities.
- Use of technology to automate routine tasks.
- Regular training in time management skills.
- Strong communication and coordination mechanisms.
- Leadership commitment to efficiency goals.
- Ongoing evaluation and process improvement.

Measuring and Evaluating Time-Based Performance

Measuring time-related performance is essential for improving operational efficiency. Organizations use metrics such as cycle time, turnaround time, and on-time delivery rates to evaluate how effectively time is managed. These measurements provide insights into process performance and help identify areas requiring improvement. Regular evaluation ensures that time management practices remain aligned with organizational objectives and adapt to changing conditions. Transparent performance measurement also promotes accountability and motivates teams to continuously improve their efficiency.

Table: Time Management Practices across Functions

Function	Time Management Focus	Impact on Operational Efficiency
Operations	Process scheduling and coordination	Reduced delays and smoother workflows
Human Resources	Workforce planning and shift management	Optimal utilization of labor
Marketing	Campaign timelines and deadlines	Timely market responses
Finance	Reporting and closing schedules	Faster decision-making
Customer Service	Response and resolution time	Improved customer satisfaction

Challenges in Achieving Time Efficiency

Organizations often face challenges such as unpredictable workloads, interruptions, and resistance to change when attempting to improve time management. Poor communication, unclear responsibilities, and inadequate tools can further undermine efficiency. Balancing speed with quality is another common difficulty, as excessive focus on time reduction may compromise outcomes. Addressing these challenges requires a balanced approach that integrates planning, flexibility, and continuous learning. Organizations that acknowledge and proactively manage these obstacles are more likely to achieve sustainable improvements.